PENGUIN BOOKS

METALS IN THE SERVICE OF MAN

Dr Arthur Street and Professor William Alexander graduated in the early 1930s from Birmingham University, where they studied metallurgy and conducted research. Until his retirement in 1975, Dr Street was chairman of a well-known diecasting company. He wrote one of the first books about diecasting, and has recently produced two modern ones. In 1975 he became the first European to win the Doehler Award, given annually by the American Die Casting Institute. His principal hobbies are grand opera, foreign travel, lecturing on art, metals and music, and preparing new editions of this book.

For many years after obtaining his doctorate the late Professor Alexander was involved in metallurgical research and production in one of the largest non-ferrous metal organizations in Britain, where he worked on copper, aluminium, niobium, beryllium, thorium and uranium, and their alloys. During the Second World War he solved in six weeks the problem of cracking in submarine periscopes which had been baffling metallurgists for many years. From 1967 until his retirement in 1976 he was Professor of Metallurgy at the University of Aston in Birmingham; he was then appointed professor emeritus, continuing metallurgical consulting and research on safety problems and the conservation of materials. Professor Alexander was president of the famous athletic club the Birchfield Harriers.

Metals in the Service of Man was first published in 1944. This tenth edition appears in the book's fiftieth year.

METALS IN THE SERVICE OF MAN

Tenth, Golden Jubilee Edition

Arthur Street

William Alexander

PENGUIN BOOKS

PENGUIN BOOKS

Published by the Penguin Group
Penguin Books Ltd, 27 Wrights Lane, London W8 5TZ, England
Penguin Books USA Inc., 375 Hudson Street, New York, New York 10014, USA
Penguin Books Australia Ltd, Ringwood, Victoria, Australia
Penguin Books Canada Ltd, 10 Alcorn Avenue, Toronto, Ontario, Canada M4V 3B2
Penguin Books (NZ) Ltd, 182–190 Wairau Road, Auckland 10, New Zealand

Penguin Books Ltd, Registered Offices: Harmondsworth, Middlesex, England

First published in Pelican Books 1944
Second edition 1951
Third edition 1954
Fourth edition 1962
Reprinted with revisions 1964
Reprinted with revisions 1968
Reprinted with revisions in Penguin Books 1990
Tenth edition 1994
10 9 8 7 6 5 4 3 2 1

Fifth edition 1972
Sixth edition 1976
Seventh edition 1979
Eighth edition 1982
Reprinted with revisions 1985
Ninth edition 1989

Copyright 1944, 1951, 1954 by Arthur Street and William Alexander
Copyright © Arthur Street and William Alexander, 1962, 1964, 1968, 1972, 1976,
1979, 1982, 1985, 1989, 1990, 1994
All rights reserved

Typeset by Datix International Limited, Bungay, Suffolk
Printed in England by Clays Ltd, St Ives plc
Set in Monophoto Plantin

With happy memories of Brenda and many thanks to Kath.
We are grateful to both of them for their help,
encouragement and patience during the preparation
of all the editions of this book.

CONTENTS

Preface ix
'Dramatis Personae' xi
List of Plates xiii
A Note on Metrication xv

1. Metals and Civilization 1
2. How We Get Our Metals 7
3. Making Iron 17
4. Making Aluminium 30
5. Alloys 37
6. Metals under the Microscope 44
7. The Inner Structure of Metals 56
8. Shaping Metals 64
9. Coinage 92
10. Testing Metals 96
11. Iron and Steel 113
12. Modern Steelmaking 118
13. The Role of Carbon in Steel 128
14. Alloy Steels 140
15. Cast Iron 146
16. Aluminium 151
17. Copper 167
18. Three Common Metals: Tin, Zinc and Lead 175
19. Nickel 187
20. Magnesium 194
21. Titanium 198
22. Some Minor Metals 201
23. Corrosion 226
24. Joining Metals 235
25. Powder Metallurgy 246

26. Metals and Nuclear Energy   250
27. The Future of Metals   266
28. Some Competitors of Metals   281
29. Metals and Materials as a Career   289

Glossary   296
Index   301

PREFACE TO THE TENTH EDITION

The first edition of this book was written in an earlier age – the 1940s – and was published by Penguin in 1944. During the war many people in factories, the armed forces and research establishments were busy making, shaping and using metals, so we hoped that the book would be helpful to them and would interest the ordinary reader too. As a result, many young people became interested in the science of metals. Indeed some, now eminent in various fields of metallurgy, have told us that they read *Metals in the Service of Man* one or even two generations ago. I hope that the final chapter will be helpful to other young men and women who are considering metals and materials as a career.

Discoveries and developments in science and technology are continuous and rapid, so every time the book became due for reprinting it was necessary to make revisions. Important uses were discovered for some metals which were merely mentioned in the earlier editions. New processes were developed: the iron and steel industry, for example, has changed beyond recognition. Ingenious instrumentation has brought us close to 'seeing' individual atoms and following their dances. Each new probe into the 'inner space' of materials can widen our use of them; each new process can improve our economy.

However, while these developments have been going on it has become increasingly difficult to win metals from their ores as the richer deposits have been depleted. More energy is required to extract metals from poorer ores, so the conservation of materials and the economical use of energy have become of vital importance.

My lifelong friend and co-author has, alas, recently died. Professor 'Bill' Alexander was well known in Britain and elsewhere as a distinguished metallurgist, with much influence also in education and sport. When Penguin Books invited us to prepare this jubilee edition, he and I decided that more emphasis should be given to non-metallic and composite materials (arguably one of the greatest developments of the second half of this century). We agreed to write more about the conservation of both materials and energy, and to indicate how, thanks to the silicon chip, the programming of manufacturing processes has been revolutionized.

I am delighted to report that Professor Alexander's family have initiated 'The Alexander Prize'. This will be awarded each year to a student in the Department of Metallurgy and Materials at the University of Birmingham, where he and I were students in the 1930s. A similar prize or prizes will be available for pupils in the King Edward's Foundation Schools in Birmingham. For many years he was a Governor of these schools and had a great interest in and influence on their activities.

It may be gathered from the preceding paragraphs that 'a little help from our friends' made it possible to revise the book every few years and remain up to date. Such help has been specially appreciated during the preparation of this jubilee edition. The staffs of metallurgical companies, research establishments, universities and development associations, and specialists in the science of materials, have kindly supplied information. Readers have written with suggestions for improvements – and sometimes with corrections.

We have always been grateful to our friend Dr Francis Fox, now in Australia, who gave us continual advice when we prepared the early editions and who has been a source of guidance for the present edition. We remember our artist friend, the late Francis Coudrill, some of whose original drawings have been reproduced in this edition (pages 97 and 269). I must record our thanks to the then Mr Allen Lane who, in the early 1940s, patiently encouraged Bill Alexander and me to embark on the first edition of this book. Finally, I thank the present staff of Penguin, who showed equal patience and encouragement in the early 1990s.

Arthur Street

'DRAMATIS PERSONAE'

Principals

IRON	The most important metal
ALUMINIUM	The light metal, second in importance to iron
COPPER	The conductivity metal
ZINC	The galvanizing metal
LEAD	The battery metal
TIN	The metal that tins the can
NICKEL	The versatile metal

Supporting characters

MAGNESIUM, BERYLLIUM	The ultra-light duet
TITANIUM	The strong middleweight
CHROMIUM	The stainless metal
TUNGSTEN	The dart-players' 'light' heavyweight
MANGANESE, VANADIUM	The scavenging metals
BORON, COBALT, MOLYBDENUM	A non-metal and two metals that give improved properties to steel
NIOBIUM (alias COLUMBIUM), ZIRCONIUM	Two important new arrivals
CADMIUM	The weather-resister
TANTALUM	The capacitor metal
GOLD, SILVER, PLATINUM	The precious trio
IRIDIUM, PALLADIUM, RHODIUM, RUTHENIUM	The valuable quartet
BARIUM, CAESIUM, CALCIUM, POTASSIUM, SODIUM, STRONTIUM	The reactive sextet

OSMIUM	The heaviest metal
LITHIUM	The lightest metal
MERCURY	The liquid metal
GALLIUM, HAFNIUM, INDIUM, RHENIUM, RUBIDIUM, SCANDIUM, THALLIUM, YTTRIUM	Some rare metals
CERIUM, DYSPROSIUM, ERBIUM, EUROPIUM, GADOLINIUM, HOLMIUM, LANTHANUM, LUTETIUM, NEODYMIUM, PRASEODYMIUM, PROMETHIUM, SAMARIUM, TERBIUM, THULIUM, YTTERBIUM	The fifteen 'rare earth' metals, known as lanthanides
TECHNETIUM	A synthetic metal
PLUTONIUM, RADIUM, THORIUM, URANIUM	Radioactive metals
ACTINIUM, AMERICIUM, BERKELIUM, CALIFORNIUM, CURIUM, EINSTEINIUM, FERMIUM, LAWRENCIUM, MENDELEVIUM, NEPTUNIUM, NOBELIUM, PROTACTINIUM (The derivation of most of these characters' names will be apparent)	Very rare metals, most of them synthetic products of radioactivity. They are classed as actinides, because the one with the lowest atomic number is actinium. This group also contains uranium, thorium and plutonium, which play a more active part than the other actinides
FRANCIUM, POLONIUM	Unstable metals, formed in the disintegration of radioactive elements
ANTIMONY, ARSENIC, BISMUTH, TELLURIUM	Semimetals or metalloids
GERMANIUM, SELENIUM, SILICON	Semiconductor elements
NITROGEN	A gas that helps to harden some metals
CARBON, OXYGEN	Two non-metals, vital to life and useful in metallurgy

LIST OF PLATES

1. Aerial view of the Bingham Canyon copper mine – the world's biggest quarry
2. The blast furnace at Redcar
3. The control room of the Redcar blast furnace
4. The 'Diane de Gabies', an aluminium statuette cast in the mid-nineteenth century
5a. A second-stage vane segment for the Pratt & Whitney F100–PW–200 fighter engine
5b. Cutaways of blades for two jet engines: (left) a single-crystal casting for the Pratt & Whitney PW 2037 engine; (right) a directionally solidified component for the General Electric Company's F 404 engine
6. A Japanese electron microscope capable of magnifying more than a million times
7. A scanning Auger microscope
8. The bubble raft experiment
9. Nickel–aluminium bronze propeller for the oil carrier STS *Hellespont Paramount*
10. Casting a marine propeller: the mould in the course of construction
11. The largest press of its kind in the UK: a 6000 tonne mechanical press, capable of producing forgings up to 60 kg in weight
12. A 4-high mill for cold rolling aluminium strip 2 metres wide, in operation at Commonwealth Aluminum Corporation, USA
13. A robot stacking bags of coin
14. A universal testing machine for measuring the tensile and compressive strengths of metals
15. The world's first iron bridge, over the River Severn at a place now called Ironbridge
16. The SS *Great Britain*, now at Great Western Dock, Bristol
17. Continuous casting of steel: British Steel's eight-strand machine at Lackenby
18. Jacket section of a North Sea oil production platform under construction at Highlands Fabricators, Scotland; the double-decker bus parked alongside gives an idea of its size

19. Ipsen Abar and Efco vacuum heat treatment furnaces, with capacities of $2\frac{1}{4}$ tonnes
20. 'Eros' – a classic example of aluminium casting
21. An automobile spaceframe, made from fewer than a hundred aluminium extrusions and interconnecting aluminium alloy diecast nodes, robotically welded to the car body
22. Two views of the Perestroika Bell
23. A Rolls-Royce RB 211 three-shaft turbofan
24. A lightweight mountain bike with a diecast magnesium alloy frame
25. Consumable electrode vacuum arc furnaces for melting titanium at IMI Titanium's plant in Birmingham
26. Drilling machine in the 'crossover cavern' of the Channel Tunnel
27. The Thames Barrier
28a. A drilling bit with tungsten carbide teeth
28b. Robotic arc welding
29. Fuel elements for various nuclear power stations
30a. Surface of cast antimony, showing dendritic structure
30b. Pearlite: a steel with about 0.9 per cent carbon
30c. Martensite: the structure of a hardened steel
30d. The structure of a steel which has been hardened and tempered
31a. Cuboids of antimony–tin intermetallic in a lead–tin–antimony bearing alloy
31b. Alloy of 60 per cent copper and 40 per cent silver, showing dendrites
31c. Alloy containing 88.5 per cent aluminium and 11.5 per cent silicon (unmodified)
31d. The same alloy as in Plate 31c, modified by the addition of 0.05 per cent sodium
32a. Dislocations in aluminium which has been cold worked
32b. Dislocation lines from particles of aluminium oxide in brass
32c. An iron–silicon alloy, showing slip bands
32d. The same iron–silicon alloy as in Plate 32c after annealing, showing dislocations and recrystallization

A NOTE ON METRICATION

Many readers under the age of 30 think of lengths and weights in terms of metres and kilograms, while some older people retain yards and pounds. The acceptance of metrication has been gradual and is still continuing. Early civilizations developed their own units of measurement. The Egyptian royal cubit equalled the width of four palms or 28 fingers (about 520 mm); a hundred cubits equalled a 'rod of cord'. A much larger measure was the 'river measure', which was about 4000 cubits (about 2 km). Many other ancient standards were based on human measurements, for example King Henry I decreed that the distance from the tip of his nose to the end of his thumb should be a standard of length, the yard.

As early as 1670, Gabriel Mouton, the vicar of St Paul's church at Lyon, suggested that the standard French measurement of length should be some proportion of the Earth's circumference. The situation in France was confusing. Individual provinces and cities used their own different units of measurement; there was, however, a military standard, the *toise du Châtelet* (about six feet), used to measure the height of conscripted soldiers. In 1791 a French committee agreed that one ten-millionth of the Earth's meridional quadrant, from pole to equator, should be the new standard. As it was impossible to take a measurement directly from either pole, a shorter distance was selected, from Dunkerque to Montjuich, near Barcelona, 9° to the south, and it was agreed that the new standard should be one-millionth of that distance, which was painstakingly measured over the next eight years.

In 1799 the standard metre, thus defined, was formed in platinum; it was replaced by one of platinum–iridium alloy in 1889. Subsequently very precise standards of length based on wavelengths of light were adopted, but the present definition of the metre is 'the length of the path travelled by light in vacuum during the interval of 1/299 792 458th of a second', the light being that emitted by an iodine-stabilized helium–neon laser.

In 1965 the British Government acceded to industry's request to adopt metric units, and the decision was also taken to adopt a particular version, the International System of Units (SI units). By 1971, when British currency went decimal, engineering, commerce and education were already 'thinking metric' and using the newly adopted units. Soon afterwards we changed the measurements in this book to metric, using the SI units, and giving sufficient information to help those who found it difficult to visualize the strength of metals in terms of newtons per square millimetre. In converting from one system of measurement to another, some approximations are necessary; for example, one pound is exactly 0.453 592 37 kilograms (by definition), but for conversions the approximate figure in the following list will suffice:

1 pound	equals	0.454 kilogram
1 inch	equals	25.4 millimetres (exactly)
1 gallon	equals	4.5 litres
1 ton force per square inch	equals	15.44 newtons per square millimetre (usually written as 15.44 N/mm²)

The derivation of the newton, in connection with the tensile strength of metals, is discussed on page 98. In several parts of the book, we have also given the strength of metals in tons per square inch for comparison.

TONNES OR TONS?

The 1000 kg metric tonne is equal to 2204 pounds and is therefore about 1.6 per cent lighter than the Imperial ton. Most industrial concerns now purchase metal by the tonne, although many people are still reluctant to use the word, perhaps because of its 'olde worlde' flavour. We have therefore compromised. Where a precise weight is indicated, for example in referring to current production statistics, we use tonnes. Where we mention the annual production of metal before metrication or a fact about the early use of metals, such as the amount wrought iron used in the SS *Great Britain* or of cast iron in the bridge over the River Severn, we have retained tons.

DISTANCES

For precise measurements we convert inches to millimetres with a multiplier of 25.4, or to centimetres with a multiplier of 2.54. Larger measurements are converted to metres or kilometres. Very small measurements are given in terms of the micrometre (a millionth of a metre, or a thousandth of a millimetre) or, occasionally, the angstrom unit (a ten-millionth of a millimetre). Instead of the latter, scientists now prefer the nanometre (a millionth of a millimetre); thus 10 angstrom units equal one nanometre.

TEMPERATURES

Temperatures in European industrial processes are expressed in degrees Celsius, after the Swedish scientist Anders Celsius; these temperatures are sometimes called centigrade. Older people are still learning to use the Celsius/centigrade scale in discussing cookery and the weather. The scientific unit of temperature is the kelvin (K). The zero of the kelvin scale is absolute zero, but its intervals are the same as those of the Celsius scale. A temperature expressed in degrees C is converted to degrees K by adding 273.15. Thus the boiling point of water is 373.15 K.

PRESSURES

The SI unit is the pascal (Pa), which equals one newton per square metre (N/m^2). One ton per square inch converts to 15.44 megapascals (a megapascal being one million pascals) or 15.44 newtons per square millimetre (N/mm^2).

TIME INTERVALS

A millisecond:	one thousandth of a second
A microsecond:	one millionth of a second
A nanosecond:	one thousand-millionth of a second
A picosecond:	one million-millionth of a second

METRICATION IN THE FUTURE?

So far as we are aware the hour will be divided into sixty minutes and the minute into sixty seconds for a long time to come, though the second is now defined as the duration of 9 192 631 770 periods of a particular type of radiation from the caesium-133 atom. The cricket pitch is still 22 yards, the American football pitch 100 yards. Although some drinks are now sold in half-litre cans, we guess that our local pub will be serving pints for many years to come.

1

METALS AND CIVILIZATION

Metals play an important part in our everyday lives. When we awake, we press an electric switch and current flows along a copper wire to light a bulb with a tungsten filament. If the weather is cold, an electric fire with a nickel–chromium alloy element provides heat. We wash in water that has come through copper pipes and runs out through a brass tap, plated with nickel and chromium, or maybe with gold. When we eat breakfast, we use stainless steel or silver cutlery. We may not realize that around us, in the microwave oven, the radio and the electric cooker, there are small but powerful magnets containing rare metals such as samarium and neodymium.

We drive to work in an automobile, which is about 80 per cent metal – mainly steel and aluminium, or we catch a train or bus, paying the fare with coins made of a copper–zinc–nickel alloy, or steel electroplated with copper. This is only the beginning of a day during which hundreds of metal objects may be used, whether we work in a hospital, drive a tractor, operate a computer-controlled milling machine or observe the universe through a telescope.

Some of the world's greatest structures are made almost exclusively of metal alloys. Supersonic and subsonic aircraft, the Golden Gate Bridge and the Sydney Harbour Bridge owe their strength to good design with the use of metals. The tunnels through the Rockies, through the Alps, under the English Channel and from Hong Kong to Kowloon were excavated with immense machines whose cutting teeth were made of a tungsten alloy.

The craft of metalworking began in about 6000 BC, when particles of native copper discovered in the Middle East were hammered into simple shapes and used for tools and decorative beads. In about 3500 BC copper was made by smelting. At first the metal must have been obtained accidentally by someone who lit a fire over ground containing a compound of copper. In the heat of the fire, wood-carbon reacted with this mineral and small pellets of copper were found when the fire had died down. Subsequently the process was carried out deliberately, and primitive furnaces were built in order to produce the metal.

Early metalworkers learned to make fires hot enough to melt metals in earthenware containers which the Romans called *crucibuli* and which we now call crucibles. Molten metal was poured into the cavity made by placing together the two halves of a hollowed-out clay or stone mould; the metal filled the cavity and, when solid, was found to have taken the cavity's shape. Archaeologists have discovered ancient bronze axe-heads made by casting in this manner which date back to 2000 BC.

It was discovered that copper and tin could be melted together, making an alloy – bronze – which was stronger, harder and tougher than the metals of which it was composed. The early metallurgists found that copper alloyed with one part in ten of tin was a good material for tools and weapons. If more tin was added the bronze was harder, while less tin gave a more malleable alloy. Gold, which occurs native,* was called a noble metal because it could be exposed to the atmosphere without tarnishing. This property led gold and its sister-metal, silver, to be used for jewellery and coinage. The Egyptians and other ancient peoples made exquisite decorative objects with these precious metals.

Owing to some similarities between them, it was imagined that the so-called base metals such as lead could be transformed into silver or gold. The mystic pseudo-science of alchemy emerged from China and Egypt in the third century BC; it spread to Europe and thrived until the eighteenth century. Alchemy became associated with the fascinating possibility of converting base metals into gold through the agency of a secret substance called the philosophers' stone, the elixir or the grand magistry.

While the alchemists and their wealthy and often exasperated patrons were trying all sorts of experiments to produce the philosophers' stone, the metalworkers, though using fewer incantations, were developing another wonderful process – the changing of dull, earthy minerals or 'ores' into metals by smelting them with charcoal in furnaces. They learned how to recognize those ores which could be smelted profitably and how to transform them into metal. It is not surprising that their efforts were sometimes unsuccessful. Even today, producers of metals encounter problems because of impurities in the ores which upset smelting operations or have a deleterious influence on the resulting metals. In those early days such happenings were attributed to the evil eye, to the

* That is, by itself, not bound up chemically in a mineral.

malevolent attention of hobgoblins or to Old Nick himself. The name of nickel was derived in that way, while cobalt's name originated from the German elves called *kobolds*.

In the Canon's Yeoman's story in *The Canterbury Tales*,* written in about 1387, Geoffrey Chaucer made fun of the metalworkers who had problems with their melting-pots:

> It happens, like as not
> There's an explosion and good-bye the pot!
> These metals are so violent when they split
> Our very walls can scarce stand up to it.
> Unless well-built and made of stone or lime
> Bang go the metals through them every time
> And some are driven down into the ground
> That way we used to lose them by the pound
> And some are scattered all about the floor;
> Some even jump into the roof, what's more.
> Although the devil didn't show his face
> I'm pretty sure he was about the place.

Between the Middle Ages and the Industrial Revolution the main progress in metallurgy was the building of larger and more efficient furnaces to produce metal in greater quantity. In 1740, when Dr Johnson used to drink tea with Samuel Richardson and Mr Garrick, Britain was the world's greatest producer of metals: the country was at war with Spain, and made about twenty thousand tons of iron per year (now the world makes that amount every hour). Twenty years after the beginning of Queen Victoria's reign, the manufacture of steel was becoming a major industry. In 1856 Henry Bessemer made public his process for converting large quantities of iron from the blast furnace into steel. The rapid development of the use of iron and steel throughout the civilized world was outstanding, though copper, tin, lead and zinc were also being produced. Metals in general, and steel in particular, came to be used for making bridges, railway lines, ships, guns, implements of all kinds, and, towards the end of the nineteenth century, those noisy 'horseless carriages' which chugged their way along the roads at a speed sometimes exceeding five miles per hour.

The nineteenth century must have been an exciting time for chemists and metallurgists as they came to realize that new elements were waiting to be discovered. One compound would be separated from the others in a mineral on the suspicion that it contained a previously unrecognized

* From Nevill Coghill's translation into modern English, published by Penguin.

element; the properties of the compound would be studied and many processes tried in the hope of isolating the element. At last a small amount would be produced, but often its properties would seem unimpressive because it was impure. In metals ranging from beryllium and nickel to vanadium, the first specimens extracted were brittle, and improvements in quality were not achieved for several years. The famous chemist and inventor Sir Humphry Davy who, as Edmund Clerihew Bentley was to observe, 'lived under the odium of having discovered sodium', was also responsible for isolating potassium, calcium, magnesium, barium and strontium.

Having isolated the new elements, scientists would wonder how they could be used commercially. What applications could be found for a metal which was much lighter than the others? We of the twentieth century are so accustomed to using aluminium that it is hard to imagine the feelings of those who produced the first dirty-looking specks of aluminium in the nineteenth century.

During the late nineteenth century* the *art* of metalworking was growing into the *science* of metallurgy. In 1861 Professor Henry C. Sorby of Sheffield initiated the microscopic examination of metals and thus laid the foundations of metallography. Properties of metals and alloys – their melting points, strengths, hardnesses and electrical conductivities – were studied, leading to improved control of manufacture of metal objects and a greater understanding of the behaviour of metals and alloys.

The expansion of the use of metals has been so rapid that considerably more metal has been extracted during the last 93 years than from the beginning of man's history to AD 1900. *Table 1* compares the world average annual production of some well-known metals, first during two 50-year periods, 1858–1907 and 1908–57, then for the periods 1958–70 and 1971–81, and finally for the single year of 1992.

The variety of available metals has certainly benefited humankind, but it has required much effort and ingenuity to get the best out of each metal and alloy. For example, Concorde was designed to withstand the arduous conditions of aerodynamic heating, coupled with the long life of 45 000 flying hours required of the first supersonic aircraft for airline use. Seventy-one per cent of the structure is made of a heat treated aluminium alloy containing 2.5 per cent copper, 1.5 per cent magnesium,

---

* The metallurgical beginnings of the Industrial Revolution are splendidly displayed at the Coalbrookdale open-air museum in Telford, between Birmingham and Shrewsbury, near the famous iron bridge.

Table 1. Average annual world production of some metals (millions of tonnes per year). There is also an ever-increasing yield of reclaimed metals, in particular 5 million tonnes of aluminium and 3 million tonnes of lead

	1858–1907	1908–57	1958–70	1971–81	1992
Copper	0.26	1.70	6.00	8.25	10.67
Lead	0.50	1.40	2.50	3.07	3.34
Tin	0.06	0.14	0.15	0.18	0.15
Zinc	0.30	1.40	3.30	5.82	7.60
Aluminium	0.002	0.76	8.00	12.00	14.76
Magnesium	—	0.04	0.15	0.25	0.31
Nickel	0.004	0.03	0.46	0.60	0.85
Steel	11	120	475	680	721

1.2 per cent iron and 1.0 per cent nickel. Sixteen per cent of the weight of Concorde is of high strength steel, mainly in the undercarriage. Titanium alloys, amounting to 4 per cent, are used in the engine nacelles, which have to withstand a temperature of about 400°C. The remaining 9 per cent comprises nickel alloys, plastics composites, glass and other materials.

During the last fifty years some metals have risen from laboratory curiosities to positions of importance: tantalum, titanium, niobium, molybdenum and zirconium. The requirements of nuclear energy, jet aircraft and space vehicles have provided incentives to overcome the immense problems of winning such metals from their ores and of fabricating them. Sometimes new methods have been developed to shape metals that, a generation ago, were thought to be quite intractable and of negligible commercial value.

A rigid classification of the elements into metals such as iron and copper, versus non-metals such as carbon and phosphorus, does not tell the whole story. Elements such as antimony and arsenic have properties characteristic of non-metals, with some metallic tendencies; they are known as semimetals or metalloids. Another group, the semiconductors, includes silicon and germanium; they have low electrical conductivities, but if small amounts (less than 1 per cent) of such materials as phosphorus are added their electrical conductivity increases dramatically, up to a millionfold. This phenomenon led to the development of transistors, which revolutionized the design of radios, computers, photocopiers and all the electronic devices which today we take for granted.

While exploitation of new metals is proceeding, continual efforts are made to widen the scope of the well-known metals. New ways of producing iron and steel; new methods of increasing the purity of metals to reveal unsuspected properties; new alloys; new methods of shaping metals and improving their strength, hardness and corrosion resistance; these are just a few of the developments which have taken place in the last fifty years. In these and other ways designers and engineers are being offered an ever wider range of metallic materials. Using the knowledge and advice of metallurgists, they are able to select the most suitable metal for the job and the best treatment for that metal, bearing in mind its mechanical and physical properties, ease of fabrication, availability and cost.

We are so familiar with metals that the difficulty of defining a metal may be overlooked. Indeed, different branches of science have different definitions. A chemist might say, 'Metallic elements are those which possess alkaline hydroxides.' A physicist might define a metal as 'An element with a high electrical conductivity which decreases slightly with increase of temperature.' Someone who works with metals would remind us that they possess lustre when polished and are good conductors of electricity and heat; and would add that most of them are denser, stronger, and more malleable than non-metallic elements. Engineers will tell you that metals and alloys have great strength and capability to withstand limited overloading without catastrophic failure – in other words, they are tough.

We, the authors, having handled a variety of metals and alloys for over fifty years, can tell you of our abiding life interest. We can reassure those following that there is much scope for craft and ingenuity in metallurgy, and that fundamental understanding of many of the properties of metals still calls for intensive scientific work.

2

HOW WE GET OUR METALS

When Commander Robert E. Peary was exploring Greenland in 1894, an Inuit (Eskimo) took him to a place near Cape York where he found three metallic meteorite masses. The Inuit called them Saviksue, or 'The Great Irons', each having a name suggested by its shape and related to the Inuit legend about how the three meteorites came to Earth. Peary removed 'The Woman' and 'The Dog' to his ship in 1895, and brought them to the American Museum of Natural History. The largest meteorite, 'The Tent', was difficult to remove and had to remain there till the summer of 1897, when Peary returned with heavy lifting gear. His four-year-old daughter, Marie Ahnighito, threw a bottle of champagne over the meteorite, which was then renamed Ahnighito. Today it is the centrepiece of the Arthur Ross Hall of Meteorites, accompanied by 'The Woman' and 'The Dog', in the American Museum of Natural History in New York.

The Hoba West meteorite, even larger than Ahnighito, was found in 1920 near the north Namibian town of Grootfontein. This monster, half-buried in the ground, has an estimated weight of 66 tonnes. Such meteorites consist of an alloy of iron with about 8 per cent nickel. No doubt primitive man, whose local culture was thus by accident raised from the Stone Age to the Iron Age, thought that meteorites were gifts from the gods. Nowadays, of course, they are not regarded as a useful source of iron: the delivery service is erratic and the unheralded arrival of a meteorite in one's back garden would be scientifically rather than commercially profitable.

Apart from metals so delivered from outer space, only a few are found native in the earth. Gold had been discovered and used by 4000 BC, when powerful rulers like those in Egypt could afford the manpower to separate grains of gold from alluvial sands and gravels. It is believed that the legend of Jason and the Argonauts' quest for the Golden Fleece has its basis in voyages in search of gold. In those days the Colchian peoples at the eastern end of the Black Sea collected gold dust from the River Phasis by using sheepskins: the metal was retained in the fleece, and the sand washed away. Many generations later, following the discovery of gold in the Sierra Nevada, the 'Fortyniner', with his daughter Clementine, panned gold from the streams of California. Nowadays gold is extracted from deep mines in South Africa by chemically separating it from the earthy material. Iron, copper and most other metals are obtained by smelting ores which contain compounds of the metals.

METALLIC ORES

The Earth's crust holds a miscellany of minerals – rocks, sand, granite, limestone, clay, volcanic lava and many others. Some contain compounds from which metals can be extracted economically, and it is necessary to discover where large ore deposits are situated and how they can be converted into metals. One would hardly imagine that strong and bright metals could be produced from such uninteresting-looking earthy substances. Yet it is easy to realize that rust is a chemical compound of iron, while verdigris is a compound of copper; these are the kinds of metal-bearing material which mineral deposits contain. The ores are treated by fire or by chemical or electrical processes to convert them into metals, these operations being known as smelting. Before that can be done, the ore deposits must be mined or quarried, concentrated, and transported to the smelting works.

The Earth's crust provides the ores from which metals are smelted, so it is interesting to discover how much of each element it contains. In 1924 two American geologists, F.W. Clarke and H.S. Washington, attempted to calculate the proportions of the elements in the Earth's crust. They took a series of statistical averages based on 5159 chemical analyses of minerals from all over the world. The result of their investigation is illustrated in *Fig. 1*.

Oxygen and silicon, which are present in sandstones, granites and most rocks, make up the major part of the Earth's crust. Compounds of aluminium, iron and magnesium exist in large quantities. Ores of some comparatively unfamiliar metals occur in considerable amounts – for example, titanium comprises one part in 160. On the other hand copper, tin, lead, nickel and mercury are among the rarest elements in the Earth's crust: copper is present to the extent of only one part in ten thousand, and lead one part in fifty thousand.

Clarke and Washington did not fully consider the relative abundance of different rock types, so the exercise has been repeated by American,

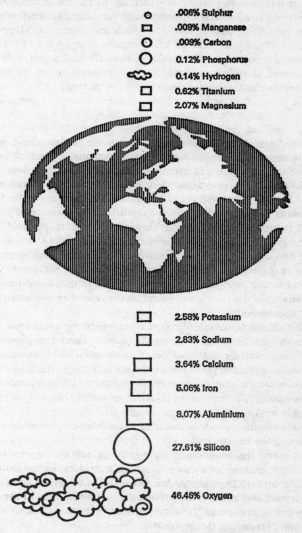

.006% Sulphur
.009% Manganese
.009% Carbon
0.12% Phosphorus
0.14% Hydrogen
0.62% Titanium
2.07% Magnesium

2.58% Potassium
2.83% Sodium
3.64% Calcium
5.06% Iron
8.07% Aluminium
27.61% Silicon
46.46% Oxygen

Fig. 1. Analysis of the Earth's crust by Clarke and Washington

German and Chinese geologists, taking advantage of the increased knowledge of the last seventy years. Nevertheless, the figures obtained in 1924 are remarkably close to more recent ones; for example, Professor Karl Wedepohl has given oxygen as 47.25 per cent, silicon 30.54 per cent, aluminium 7.83 per cent and iron 3.54 per cent.

Most useful ores contain the metal combined with oxygen, sulphur or other elements, as shown in *Table 2*. Pure minerals are rarely found in nature; they are generally associated with earth, sand, limestone, clay, gravel and stones, this unwanted material being known as 'gangue'. The availability of metals depends on the ease with which their ores can be mined and smelted. An ore deposit near Salt Lake City or Kiev is likely to be more useful than similar ones below the ice of Greenland or at the bottom of the Pacific Ocean under the Mindanao Trench. The large deposits of iron ore in Brazil are estimated to be 50 000 million tonnes, but because of their remoteness they have begun to be exploited only recently.

A customer of the local inn would doubtless agree that half a pint in one glass is preferable to sixty teaspoonfuls in sixty glasses, and an analogy can be made with the occurrence of metallic ores. A single rich, extensive ore deposit of a comparatively scarce metal can be exploited more readily than many small pockets spread all over the globe. Ores of copper, lead, zinc and nickel form only a minute proportion of the Earth's crust, but they are often found concentrated in deposits that can be mined economically.

Although minerals containing aluminium make up a large proportion of the Earth's crust, some compounds of this metal are unsuitable as ores. Clay contains about 25 per cent* of aluminium in chemical combination with silicon, oxygen, iron, calcium and magnesium. A hundred barrow-loads would provide enough aluminium to make a small aircraft but, for as long as rich deposits of bauxite, the principal ore of aluminium, are available in Australia, Guinea, Jamaica and elsewhere, the expensive refining and smelting processes that are necessary to extract aluminium from clay will not be required.

It takes about ten years from the beginning of exploration of an ore deposit to the start-up of a mine. Geological studies, reconnaissance of the area and surveys from the air are followed by test drillings to confirm that the deposit will be economic. Before the mine can be developed it is necessary to build roads and railways, airports and towns, and to ascertain that enough workers will be attracted to live in the area.

* In this book, all percentage compositions are by weight, unless stated otherwise.

Table 2. Some minerals from which metals are obtained

| Metal | Name of mineral | Metallic compound contained in mineral | |
		Chemical name	Chemical formula
Aluminium	Bauxite	Hydrated aluminium oxide	$Al_2O_3 . 3H_2O$
Copper	Copper pyrites	Copper–iron sulphide	$CuFeS_2$
Iron	Hematite	Iron oxide	Fe_2O_3
	Magnetite	Iron oxide	Fe_3O_4
Lead	Galena	Lead sulphide	PbS
Magnesium	Magnesite	Magnesium carbonate	$MgCO_3$
	Dolomite	Magnesium–calcium carbonate	$MgCO_3 . CaCo_3$
Mercury	Cinnabar	Mercury sulphide	HgS
Nickel	Pentlandite	Nickel–iron sulphide	$(Fe,Ni)_9S_8$
Silver	Argentite	Silver sulphide	Ag_2S
Tin	Cassiterite	Tin dioxide	SnO_2
Titanium	Rutile	Titanium oxide	TiO_2
Zinc	Zinc blende	Zinc sulphide	ZnS
	Calamine	Zinc carbonate	$ZnCO_3$

The locations of ore deposits, called placements, are distributed irregularly over the Earth. There are some places where Nature has been lavish. Australia has been described as an Aladdin's cave of metal ores. The Hamersley Iron Province, near Port Hedland, contains massive deposits of high grade iron ore. In South Australia there are older orefields, with splendid names such as 'Iron Knob', 'Iron Baron' and 'Iron Queen'. The continent also has major sources of lead, nickel, gold, titanium, zinc and other metals. Furthermore, in the last decade immense deposits of the magnesium ore magnesite have been developed at Kunwarara in Queensland, only 100 km from the city of Gladstone, an international port. This will bring Australia into the first division of magnesium producers before the end of the century.

The industrial power of countries depends very much on the metal ores they possess. *Table 3* shows how 27 countries rank according to the amounts of metal ores which they mine. There are a few countries – the USA, Australia, Canada, China, South Africa and the countries of the former USSR – which each mine over a fifth of the total world

Table 3. The major suppliers of the world's metal ores

Country	First division: over 20 per cent	Second division: 10–19 per cent	Third division: 5–9 per cent
Albania			Chromium
Australia	Aluminium	Iron, lead, titanium	Nickel, silver, magnesium
Bolivia			tin
Brazil		Iron, manganese, tin	Aluminium
Canada	Uranium, nickel, titanium	Zinc, silver	Copper, lead, magnesium, molybdenum, gold
Chile		Copper	Molybdenum
China	Tungsten	Lead, iron, manganese	Gold, zinc
France			Uranium
Gabon		Manganese	
Guinea		Aluminium	
India		Chromium	
Indonesia		Tin	Nickel
Jamaica		Aluminium	
Malaysia		Tin	
Mexico		Silver	
Namibia		Uranium	
New Caledonia		Nickel	
Nigeria			Uranium
North Korea			Tungsten
Norway		Magnesium	Titanium
Peru		Silver	Zinc
Portugal			Tungsten
Russia*	Iron, chromium, manganese, nickel	Copper, gold, lead, vanadium, tungsten, silver, tin, molybdenum, zinc	Aluminium
South Africa	Gold, chromium	Manganese, vanadium	
USA	Magnesium, molybdenum	Copper, gold, lead, zinc, vanadium	Silver, titanium
Zambia			Copper
Zimbabwe			Chromium

* And other countries of the former USSR.

production of some metals, but many other countries are important in the 'second division'. For example, Gabon is a major producer of manganese ore, and Namibia has large uranium ore mines.

Some metals are used mainly as chemical compounds. For example, in spite of the ever-increasing uses of metallic titanium and its alloys, the biggest tonnage is represented by titanium dioxide, which is used as a pigment. Immense amounts of manganese are required by the steel industry, but the metal itself is rarely manufactured. It is produced as ferro-manganese by smelting mixed ores of iron and manganese; in the USA 6 kg of ferro-manganese is required per tonne of steel.

The economic value of an ore depends on the market price of the metal and its ease of exploitation. An iron ore containing less than 50 per cent metal is reckoned as poor, and at present would not be worth smelting. In contrast, one part of gold in 100 000 parts of ore is well worth exploitation. Vaal Reefs, a large South African mine, treated 11.5 million tonnes of ore in one 12-month period in the late 1980s to win only 80 tonnes of gold. The percentage of metal in ores varies widely: for example, copper ores yield an average of 0.5 per cent metal in the USA, 0.6 per cent in Canada and 1.0 per cent in Australia; the world average is 0.8 per cent. But the Neves Corvo mine in Portugal, unfortunately not a large one, provides copper ores with 9.3 per cent metal, the richest in the world.

OBTAINING THE ORE

Many ores are recovered from the earth by deep mining in which a shaft is sunk and the ore is blasted out in underground galleries and hauled to the surface. Gold mines are by far the deepest. East Rand Proprietary mines have gone down to 3474 metres, while at Western Deep Levels, the world's deepest mine, gold has been removed at 3582 metres below the surface. At such great depths the technical problems and those of human endurance are immense. The rock temperature is around 52°C (125°F) and, to achieve an acceptable working temperature of 27°C, large quantities of chilled water are provided from refrigeration plants situated underground or at the surface and then distributed through a pipe network. Efforts to cool the mines are now being taken a stage further by circulating a slurry of water and ice. In large mines more than 70 000 cubic metres of cold air are required per minute.

Metallic ores on or near the surface are removed by quarrying on a scale which makes one realize how much we depend on earth-moving machinery. The world's largest mine, at Bingham Canyon, near Salt

Lake City, is four kilometres across and nearly a kilometre deep. Every year sufficient ore to make 180 000 tonnes of copper is excavated and smelted. About 225 000 tonnes of overburden and ore are removed each day, the ore containing 0.7 per cent copper in the form of copper–iron sulphide. *Plate 1* shows an aerial view of the mine; the concentric rings are the terraces which have been excavated. The work is done with the aid of twelve huge electric shovels, ten drilling machines, several trains, and 49 trucks each with a capacity of about 180 tonnes. The ore is transported from the mine to the concentrator on an 8 km long, 2 m wide conveyor belt which travels at over 250 m per minute. At the concentrator the unwanted material is separated from the sulphide mineral, leaving a concentrate containing about 28 per cent copper. This is pumped from the plant in the form of a slurry, which flows 30 km to the smelting plant, where the ore is smelted to copper.

Many different types of transport are used to convey ores from mine to processing plant. Where the system needs to be flexible over short distances, trucks, often with capacities of 150 tonnes, are used. Conveyor belts, sometimes over 20 km long, carry some minerals. Trains can be the most efficient method in remote areas. The iron ore mined at Paraburdoo in Western Australia is transported nearly 400 km to the port of Dampier in trains 1½ km long, carrying 20 000 tonnes of ore. As at Bingham Canyon, some ores are formed into a slurry and conveyed in pipelines. One pipeline 165 km long conveys concentrated copper in ore slurry from the Escondida mine, situated 3100 m above sea level in Chile, to the port of Coloso.

Some metallic ores occur as waterborne deposits, and methods based on the 'panning' of the old-time gold prospector are employed. A more up-to-date form of panning has been used in the treatment of tin ores in Malaysia, where huge dredgers dig up tin-bearing gravel from lake beds and pass it over long troughs fitted with shallow projections which retain the dense ore.

CONCENTRATING THE ORE

Iron ores contain 10 to 30 per cent earth, sand and other unwanted materials. These contain silica, which when melted becomes a viscous liquid that tends to combine with iron oxide and other metallic compounds. To prevent this from happening, limestone is mixed with the iron ore and, in the heat of the blast furnace, it combines with the silica to form a fluid slag which can then be removed.

Magnesium and aluminium are made by electrical processes which will not operate unless a very pure metallic compound is separated first. A chemical method of concentrating aluminium ores is described in Chapter 4. Ores of copper, zinc, tin and lead contain only a small amount of metal, so the unwanted material must be removed before smelting. Some weak ores are concentrated by passing them, with a stream of water, over vibrating metal tables. This causes the light earthy material to move in one direction and the heavier metallic compounds to be delivered to another outlet.

In contrast to this, the widely used flotation process causes the dense metal compounds to float in a bath of frothed liquid while the unwanted minerals sink. A story is told about the discovery of this process. In a certain lead-mining village, a woman was washing her husband's working clothes in a tub in her backyard. The water was full of suds and the clothes were impregnated with particles of galena, the lead ore which he mined. The peculiar thing about this wash-tub was that the soapsuds, instead of appearing as a white foam, were distinctly dark. Some observant fellow noticed that the surface of the suds was covered with tiny particles of galena. This was surprising, for the lead ore, being very dense, would be expected to fall to the bottom of the tub and not be floating on the surface of the water. The anonymous observer saw in this wash-tub the germ of a commercial process for separating finely crushed lead ore from the gangue associated with it. Americans who tell this story maintain that it occurred in an American mining village; British narrators insist that it was in Derbyshire.

The development of flotation ranks in importance with the discovery of smelting. Virtually the entire world's supplies of copper, lead, zinc and silver ores are first collected in the froth of the flotation process. For flotation to take place, air bubbles must attach themselves to particles of metallic compounds and lift them to the water surface. Flotation is a continuous process, with thirty or more tanks arranged in line, each containing water, oil and frothing agents. Air is pumped to each tank to form the froth. After grinding, the finely crushed mineral is fed into the first tank; the earthy unwanted material sinks but the metal-containing particles are captured by the froth, whence they are skimmed off and dried. The overflow passes to the next tank, where more of the metallic compound is captured by the froth, and so on down the line.

FROM ORE TO METAL

As supplies of rich metallic ores will not last for much longer, worldwide research has been devoted to the extraction of metals from weak ores or even from the depleted tailings dumped as residues from previous extraction processes. During the past forty years, solvent extraction has become of immense importance. It has been predicted that by the end of the century the revolutionary impact of solvent extraction will be comparable to that of froth flotation at the beginning of the century.

The process depends on a complex series of reactions in which organic chemicals dissolve the metal compounds in a mineral. The solution is then separated and goes on for electrolytic extraction of the metal; the organic solvent is reclaimed. At first, solvent extraction was used to separate less common metals such as zirconium, hafnium, niobium and uranium. It also proved to be a satisfactory method of separating metals having similar chemical properties, such as niobium from tantalum, cobalt from nickel, and the various rare earth metals (page 215) from one another.

Nowadays the process is widely used to extract copper from crushed ore which has too low a concentration of the metal for economic extraction by other means, and from waste heaps that have accumulated from earlier mining. Plants are operating in the USA, Canada, Australia, Peru, Chile and Zaire. Zambia has the largest operation, with an annual capactiy of 120 000 tonnes of copper. Although at first the quality left something to be desired, recent improvements in the technology have ensured that the copper produced by solvent extraction meets the high standards required by modern users.

Solvent extraction has also been used to obtain copper from ore bodies without any conventional mining taking place. A series of wells is sunk and acid is pumped through the rock to dissolve the copper compound. Initial results have indicated that such a process causes less disruption to the environment than ordinary mining, enabling the surroundings to be easily restored to their original state when all the copper mineral has been removed.

Biotechnology is opening new fields for metal extraction. Metal sulphide minerals are leached and interacted with bacteria which convert the sulphides into oxides. Often this process is followed by solvent extraction, and finally the metal is extracted from the enriched ore. This method of metal recovery using micro-organisms is showing promise for the treatment of weak ores of copper, nickel, cobalt and uranium.

The smelting of the two best-known metals is described in the following chapters. The production of zinc is mentioned on page 178, nickel on page 188, magnesium on page 197 and titanium on page 198.

3

MAKING IRON

The production of iron began in the second millennium BC, when by chance a wood fire burnt on ground containing a compound of iron. The carbon in the wood reacted with the compound, and when the fire died down there was left a spongy mass of iron which could be hammered with a stone to form it into a weapon or tool. The chemical reaction in the fire could be expressed by the formula

iron oxide + carbon ⟶ iron + carbon dioxide
(mineral) (in the wood) (metal) (gas)

After such fortuitous discoveries our metallurgical forefathers realized that some special minerals could be converted into a useful, hard metal. They built primitive furnaces to make iron, and increased the supply of air by fanning the flames or using bellows, so that the metal could be made more rapidly than before.

During the fifteenth century AD, furnaces with a blast of air were built in Britain to produce a form of iron that could be poured from them. This metal contained about 4 per cent of carbon and some impurities. Production was wasteful because much of the iron compounds in the mineral combined with silica in the earthy material and thus was not converted to metal. Eventually it was found that limestone added to the contents of the furnace would combine with the earthy material to form a molten slag which, being lighter than the molten iron, could be run off separately.

The cast iron produced in those early blast furnaces contained carbon and melted at a lower temperature than the pure metal – about 1200°C, compared with 1532°C. It was rather brittle but could be changed into practically pure iron by a process called 'puddling' (described on page 114). This wrought iron could be hammered into useful shapes while red hot. At that time, most of the ironmaking furnaces had four or five puddling furnaces near by.

At first, charcoal obtained from wood was used for smelting iron. One furnace, built in 1711 at Backbarrow, near Lake Windermere, was 6 metres high and 2½ metres square. It was built into the side of a hill, so that the workers could tip their baskets of ore or fuel into the top of the furnace from a platform. The work was done by seven men, aided by their families. Today the works lie derelict, but the area is being surveyed and it is hoped that eventually the site will be restored. At Duddon in the Lake District National Park a completely restored site has been presented in a skilful but low-key way. The Bonawe furnace in Scotland and the Dyfi 'Ffwrnais' at Eglwys-fach on the road between Machynlleth and Aberystwyth are tourist attractions. The Welsh example has a reconstructed water-wheel, similar to the one which originally provided the air blast.

When charcoal was used to smelt iron about one acre (4000 square metres) of woodland had to be cut down per ton of iron.* The destruction of forests for this purpose made it necessary to experiment with other fuels. In 1708 a young ironfounder, Abraham Darby, leased a 160-year-old furnace at Coalbrookdale on the River Severn, near Shrewsbury. A year later he succeeded in smelting iron using coke which, being a purer form of carbon than coal, introduced fewer impurities into the metal. At first the quality of his cast iron was not very good, but it was suitable for making pots and other domestic articles. After his death Darby's efforts were continued by his family and in 1750 his son, Abraham II, succeeded in making a good quality cast iron that could be used on its own or converted to wrought iron. The improvements and cost savings made possible by the Darbys' achievements provided the starting point for the manufacture of iron rails, the iron bridge over the Severn, iron ships, iron aqueducts, and ultimately the worldwide ferrous industry.

IRONMAKING TODAY

The blast furnace is the central feature of a complex facility for the production of iron. Although, by its massiveness, it dominates the scene, it is only one link in a chain of processes. There is plant to carbonize coal into coke, to charge sintered ore, limestone and coke into the furnace, to clean and handle the vast quantities of the gas discharged from the furnace and used for pre-heating the air blast, to move the hot iron to the

---

* It is deplorable that thousands of square kilometres of Amazonian forest are being destroyed because Brazilian charcoal-burning furnaces are making over 100 000 tonnes of iron per year for export to Europe, Japan and China.

oxygen steel furnaces for conversion into steel, and to prepare for sale the slag produced as a by-product.

Figure 2 illustrates the basic features of a typical large furnace, and *Plate 2* shows British Steel's No. 1 furnace at Redcar. This blast furnace, one of the largest and most efficient in the world, is capable of producing over 10 000 tonnes of iron per day.

IRON ORES

Iron forms several oxides which are available as ores. Hematite contains Fe_2O_3 which, if pure, would yield 70 per cent iron. Magnetite is a magnetic iron ore, the oxide being Fe_3O_4. Other ores contain the monoxide, FeO, and iron carbonate. The world's annual production of iron is more than 500 million tonnes, for which about 1000 million tonnes of iron ore is required. Japan, Western Europe and the countries of the former USSR produce close on 100 million tonnes of iron a year, China and Korea about 75 million tonnes, and the USA about 50 million tonnes. Japan is a large importer of iron ore, able to bargain keenly, offering long-term contracts with Brazil, India, Australia, New Zealand and various countries in Asia. The USA itself provides about two-thirds of its ore, and obtains the rest from Brazil and Canada. Most of the iron ore required by Britain comes from Australia, Brazil and Canada, with contributions from Mauritania, South Africa, Spain, Sweden and Venezuela.

In Russia, near Magnitogorsk, there is a vast deposit of magnetite, known as the Kursk Magnetic Anomaly, which contains an estimated 80 000 million tonnes of iron ore, with 45–65 per cent metal content, plus about 10 million million tonnes of poorer ore. This area covers some 200 000 km^2, and is said to contain enough ore to produce 250 million tonnes of iron a year for the next 15 000 years.

The sea transport of iron ore represents the largest dry bulk seaborne commodity. The two next largest, grain and coal, together account for less than half the weight of the iron ore trade. Economic transport requires large vessels, and there are several of more than 200 000 tonnes capacity, with some carriers in the region of 300 000 tonnes. The new breed of ore carriers has made it viable to transport ore from many parts of the world. Deep ore terminals have been constructed to permit the use of such vessels.

New Zealand was responsible for the Marcona process, an impressive development in the treatment and transport of iron ores. The west coast of North Island contains massive deposits of black ironsands, which are mined and concentrated at Taharoa, 150 km south of Auckland, for

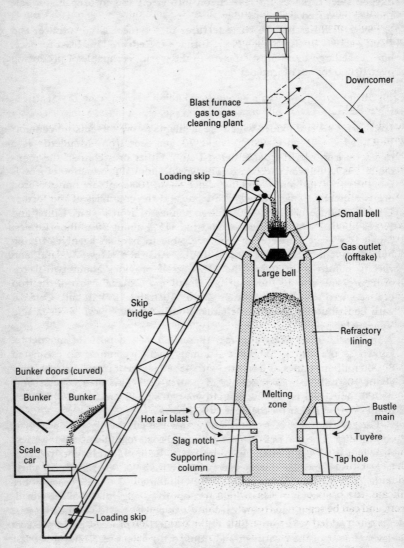

Fig. 2. A modern blast furnace (*Courtesy British Steel*)

export to the Japanese iron and steel industry. By using gravity cone separators and then wet magnetic drums, the oxide is separated from extraneous material and pumped in slurry form through pipelines to bulk ore carriers moored offshore.

COKE FOR THE BLAST FURNACE

Coals differ greatly in their properties and in the amount of volatile material that is given off when they are heated; only a minority are suitable for coke-making. Special high quality coals are selected, usually with a volatile content of between 22 and 27 per cent; strong pieces of coke are necessary to permit uniform gas flow in the furnace and the coke must be hard enough to avoid being crushed under the heavy load. The coal is heated in closed containers in the absence of air, and the temperature is gradually raised. After about 18 hours the coke is pushed from the oven, collected in a skip and quenched in water. It is then screened to remove undersize particles and conveyed to the blast furnace. In a typical plant about 30 000 tonnes of coal is carbonized to produce 24 000 tonnes of coke per week. In addition to its fundamental roles of providing the source of heat and forming the reducing gas carbon monoxide, coke provides a permeable medium through which the reducing gas can pass. One of the many problems facing ironmakers today is the high cost of producing coke, and on page 28 mention will be made of the current efforts to use coal directly in the blast furnace, instead of coke.

LIMESTONE AND SLAG

The iron ore, containing 60–80 per cent iron oxide, is associated with sand and other earthy materials, known as gangue. If the ore were smelted alone in the blast furnace a great deal of the iron would combine chemically with silica in the sand and the other materials in the gangue; this is what happened in primitive iron furnaces. Limestone – calcium carbonate – has the property of combining with the silica. In the heat of the blast furnace it is decomposed to calcium oxide, CaO, and carbon dioxide. The calcium oxide combines with the silica to form calcium silicate, the basis of the slag, which has a lower melting point than the iron, and can be separated from it. Dolomite (magnesium–calcium carbonate) is often added to the limestone in the proportion of about one part in three; this helps to reduce the viscosity of the slag. The slag from the blast furnace is sold as aggregate for road construction, railway ballast and other purposes, but some may have to be dumped.

SINTER

A very important factor, which has greatly increased the efficiency of the blast furnace, has been the sintering of the ore. Instead of charging irregular pieces ranging from small particles to large lumps, the ore is now sintered to form pieces between 10 and 25 mm. The ore is mixed with limestone and coke 'fines' to provide combustion. The materials are weighed and proportioned automatically, and mixed in a drum, and water is added to agglomerate the mix. This mix is put into pallets which are then sealed from the atmosphere and heated to a temperature of about 1200°C. The fused material so produced is crushed, screened and cooled, ready for charging into the furnace. This ensures that the contents of the furnace are distributed evenly, permitting the reducing gases of the blast to rise regularly and rapidly.

CHARGING THE FURNACE

The sintered ore, coke and limestone are given the collective name of the 'charge'. The charge is hoisted to the top of the furnace, as shown in *Fig. 2*. In some plants it is conveyed to a double bell and hopper, shown on the left in *Fig. 3*. The small bell is lowered, as shown on the right, dropping the charge into a lower chamber, at the bottom of which is a larger bell and hopper. When the top has been closed, thus sealing the furnace, the larger bell is lowered, allowing the charge to fall into the furnace. The bell is rotated through an angle each time it is used.

Very large furnaces pose problems in the operation of bell-and-hopper mechanisms: the top ring associated with bell-charged furnaces is difficult to maintain pressure-tight under the high pressure of the hot blast. A recent development to overcome these difficulties is the bell-less top, in which the charge is distributed from sealed bunkers at the furnace top by means of a rotating chute. *Figure 4* shows the bell-less top installed at the No. 1 furnace at Redcar. A full-scale replica was constructed to form a basis for the mathematical model of the process. This enabled British Steel to assess the effects of any changes in material distribution when the bell-less top came into operation.

THE HOT BLAST

The primary function of the air blast is to enable the coke to burn and produce a high temperature. Oxygen in the air combines with the coke to form carbon monoxide, which reacts with the iron oxide to produce iron

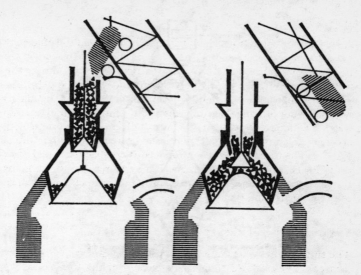

Fig. 3. Bell and hopper for charging a blast furnace

and carbon dioxide. In the early years of iron production, cold air was blown into the furnace but as much as eight tons of coal or coke were needed to produce one ton of iron. In 1828 a Scot, James Neilson, developed the use of a heated air blast, which reduced the fuel consumption to five tons per ton of iron – a great achievement at that time, though the efficiency of modern blast furnaces is about ten times better than Neilson obtained. Those who wish to pay tribute to this metallurgical pioneer can find an obelisk to his memory on a hilltop near Kirkudbright.

The hot air is blown into the furnace through a number of nozzles called tuyères. Depending on the capacity of the furnace there are between 12 and 36 tuyères around the circumference, placed sufficiently high above the hearth to leave a space in which the molten iron and slag can accumulate. Steam or oxygen may be added to the blast; several forms of hydrocarbons are sometimes injected at the tuyères – oil, natural gas and powdered coal – though not all furnaces are equipped for all these options. The main attraction of such enrichment is that the volume of the 'burden' can be reduced, so that more iron can be made in a furnace of a given size.

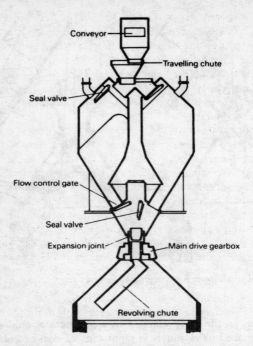

Fig. 4. Bell-less top for charging a large blast furnace

The hot gas leaving the furnace provides an excellent method of pre-heating the blast. This gas contains 20 to 25 per cent carbon monoxide, which is combustible. Some of the furnace gas is used to heat the incoming air blast by means of 'stoves', usually three to each furnace. These stoves are nearly as large as the blast furnace itself; they consist of a chamber in which the remaining carbon monoxide is burnt, and a labyrinth of firebricks arranged so that the burning gases passing through them heat them to red heat. In operation two processes are carried out at the same time in the three stoves:

1 two of the stoves are being heated, as described above;
2 the third stove, having been heated, raises the temperature of the air blast as it passes to the furnace.

By this method, the air has been raised to between 850 and 1250°C by the time it reaches the blast furnace. About every half-hour these operations are changed; the stove which was previously 'on blast' is heated, while one of the stoves which was 'on gas' now gives up its heat to the incoming blast of air. Normally the cycle is half an hour on blast and one hour on gas. As will be seen from *Fig. 5*, the gas is passed through dust extractors before being led to the stoves. The remainder of the valuable furnace gas is used for other purposes – for driving pumping engines, and for generating electricity to light the works, and often a neighbouring town as well.

The flame temperature at the tuyères is about 2200°C, and the contents of the furnace reach a temperature of about 1800°C, nearly 300°C higher than the melting point of pure iron and over 400°C higher than the melting point of iron containing about 4 per cent carbon, which is being produced in the furnace. As the process continues, iron is made from the ore and then melted by the heat of the furnace. Words like 'trickle' or 'fall' are inadequate to describe the conditions in which the enormous contents of the furnace move downwards at the rate of nearly 500 tonnes per hour, while the blast of air rushes upwards. During its descent the iron absorbs other elements from the coke, gangue and limestone, so that the pool of iron at the bottom of the furnace contains 4–5 per cent carbon, and up to 1 per cent manganese and silicon (though in some circumstances the silicon content may be only about 0.2 per cent). In some types of iron, phosphorus is present up to 2 per cent; in others, it is only a fraction of 1 per cent. There is usually a small amount of sulphur, kept as low as possible.

Until fairly recently blast furnaces were constructed with two 'tap holes' sealed with plugs of refractory material, the upper opening for the removal of slag and the lower for the hot iron, as shown in *Fig. 2*. In some modern blast furnaces the hot metal and slag are cast through the same tap hole, and separated later. This is possible in particular if a rich ore is being smelted, so that the ratio is as low as 300 kg of slag per tonne of iron. The hot iron is taken to oxygen steel furnaces (discussed in Chapter 10).

MODERN BLAST FURNACES

The typical annual output for a blast furnace two hundred years ago was 900 tons. Some modern furnaces make that amount in two hours! Such a furnace, producing 10 000 tonnes of hot iron every 24 hours, has to treat 15 000 tonnes of ore, using 4500 tonnes of coke, 14 000 tonnes of hot air in the blast, 300 tonnes of oil injected in the blast, 1000 tonnes of limestone and 200–300 tonnes of dolomite. The construction of the

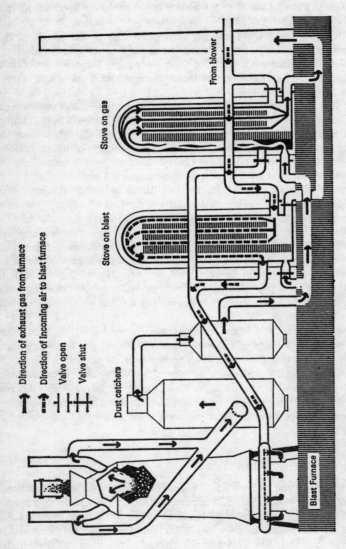

Direction of exhaust gas from furnace
Direction of incoming air to blast furnace
Valve open
Valve shut

Dust catchers

Stove on blast

Stove on gas

From blower

Blast Furnace

Fig. 5. Furnace exhaust gas is used to heat incoming air

modern blast furnace requires over 4000 tonnes of steel for the furnace body and another 4000 tonnes for piping and machinery. Japan has some special problems because of earthquakes, and the foundations of their furnaces contain 10 000 cubic metres of concrete.

An important economic factor in operating a blast furnace, and indeed in any process, is to achieve maximum output with minimum stoppage time. A major cost is incurred when the blast furnace has to be shut down for relining, made necessary by the attack of the molten iron and slag descending the furnace and the red-hot air blast hurricaning up it. Normally a furnace needs relining every six to eight years. One of the many reasons for developing very large furnaces is that doubling the inner surface area of a container approximately trebles its volume, so that over their working lifetimes larger blast furnaces should cost less per tonne of iron to reline than a smaller furnace.

An immense blast furnace would be immensely troublesome and unprofitable unless its operation were supported by a wide range of controls. Every aspect of furnace operation is recorded by a series of coordinated computers, such as those shown in *Plate 3*. They not only record the signals from the plant, but also store and assemble data for long-term information. The furnace is monitored to control the operation of the charging mechanism, and electronic probes measure the distribution of the charge and the liquid levels in the hearth. The temperature of the hot blast, the furnace exhaust gas, the refractory walls, the iron and the slag are measured and recorded. The pressure of the blast and its flow rate, the pressures at each individual tuyère and the performance of the cooling water are all fed to the control room.

During the past fifteen years the fuel consumption of blast furnaces has been reduced from about 650 kg of coke per tonne of hot iron to under 500 kg. Japan took the lead first, but then the Redcar No. 1 furnace achieved 460 kg of coke per tonne when fuelled by coke alone; with 30 kg of oil per tonne of coke injected into the blast, the coke consumption goes down to 425 kg per tonne.

NEW DEVELOPMENTS IN IRONMAKING

Every industry, large and small, needs to take a look at itself fairly frequently. Labour and fuel costs escalate; raw materials become scarce; capital costs mount; new processes are discovered. Very often the trouble is that, soon after an industry has invested a great deal of money in capital equipment, some new factors arise. If you have just spent a billion pounds sterling or one and a half billion dollars on a new iron-

and steelworks, scrapping it is painful. Blast furnaces have increased in efficiency, but coke, on which they have depended for over 250 years, is becoming more and more expensive and supplies of suitable coking coal are becoming scarce.

In Europe, Japan and the USA, much effort is being expended on developing a new generation of processes aimed at producing liquid iron using non-coking coal, in order to avoid the costs of producing coke. These efforts are a natural follow-on to the development, over the past thirty years, of various iron reduction routes, the so-called 'direct reduction processes', which remove the oxygen associated with iron in the ore to yield a solid product which is then melted in electric arc furnaces for refining into steel.

In Mexico, an increasing demand for steel made it necessary to import scrap, with resultant problems of price fluctuations. In response, the steelmakers HYLSA pioneered a direct reduction process, known as the HYL process, which enabled them to use local ores and gaseous fuel. Natural gas is mixed with superheated steam, and is converted under high pressure and a temperature of about 800°C first into methane, CH_4, and then into a mixture of carbon monoxide and hydrogen. The conversion is made possible by a nickel-based catalyst packed in stainless steel tubes. Pellets of iron ore are treated with the gaseous mixture at a temperature of about 800°C, high enough to reduce the oxide to metallic iron but not sufficient to melt the iron. The ore is converted to spongy iron pellets of high purity, suitable for conversion to steel.

Although several of the original plants are still operating successfully, there have been many improvements during the past decade. The plants are designed with two independent sections, one for generating the reducing gas and the other for converting the ore to metal. Iron ore in pellet form is transported by conveyor to the top of the reduction tower, where an automated system allows it to be fed into the top at atmospheric pressure, followed by pressurization in an intermediate section. The oxide is reduced to iron by the action of the hot reducing gases. A rotary valve at the outlet of the reactor regulates the continuous movement of the charge downwards. The iron is removed at the bottom of the reactor automatically, and is discharged at a temperature of about 50°C.

Since the process was first introduced, about 70 million tonnes of iron have been made with the HYL process. Several countries still operate successfully what is known as HYL-I, including some plants in Mexico, Venezuela, Iran, Iraq and Indonesia. The modern HYL-III process is being operated in India, Indonesia and Malaysia, and also in Mexico.

Another important process is known as Midrex, and originates in

Canada. It uses a reforming system to produce the reducing gas, and the reduction takes place in a shaft furnace about 40 metres high. The reducing gas is usually hydrogen-rich, but also contains carbon monoxide. Reduction occurs in the top half of the furnace at a temperature of about 850°C; the bottom half is an inverted truncated cone in which the iron that has been produced is cooled to near ambient temperature before being discharged. A unique feature of the Midrex furnace is its system for recycling the exhaust gas, which contains carbon dioxide: it is fed to reformers where the carbon dioxide and methane from natural gas form carbon monoxide and hydrogen.

At least fifty other processes for the direct reduction of ore to metal have been invented, but many of them have been only for scientific research. The current processes are usually viable in countries which have ample reserves of natural gas. Some plants are economic even in comparatively small units, and they are often linked with mini-steelworks in countries which are not yet highly industrialized and which therefore do not possess the vast amounts of scrap iron and steel on which the major iron- and steelmaking countries depend.

4

MAKING ALUMINIUM

Aluminium is the most plentiful metal in the Earth's crust, of which it makes up about 8 per cent. The pure metal does not occur in nature but alum, which is a hydrated potassium–aluminium sulphate, was known and used for many centuries before the metal was isolated. During the early nineteenth century attempts were made to produce the elusive element. In 1827 Friedrich Wöhler treated aluminium chloride with potassium metal, forming a grey powder consisting of aluminium heavily contaminated with oxide.

Over the following years the process was improved to the point where pieces of aluminium as big as pinheads could be obtained. This was sufficient to measure some of the properties of the metal and to discover that it had a much lower density than the other metals known at that time. In 1854 Henri Etienne Sainte-Claire Deville made a small ingot of aluminium by a chemical reaction between sodium and sodium–aluminium chloride. It is remarkable that soon afterwards an aluminium statuette was cast, about a metre high, being a copy of the Diane de Gabies in the Louvre, illustrated in *Plate 4*. This casting must have represented most of the world's annual production of aluminium!

Deville and others realized that the metal might be obtained by electrolysis of the oxide, alumina, but this compound has a high melting point, about 2000°C, so alone it would not be suitable. However, the separation could be achieved if a substance were found that would dissolve alumina, forming a solution with a much lower melting point. These were some of the problems to be solved when, in the 1880s, Charles Martin Hall in the USA and Paul Louis Toussaint Héroult in France* were experimenting with the electrolytic method, taking advantage of the great new possibilities of cheap and plentiful electric power, following Gramme Zenobe's construction of the first practical dynamo.

* Hall and Héroult were both born in 1863, and they both died in 1914.

Hall and Héroult worked separately, each unaware of the other's researches, until in 1885, after heartbreaking failures and little encouragement, the two pioneers both arrived at the same solution by using the mineral cryolite, whose chemical composition is sodium–aluminium fluoride and which, when molten, is a good conductor of electricity. They found that molten cryolite dissolves about 5 per cent alumina; the solution melts at a little under 1000°C and an electric current passed through it will keep the liquid molten and split up the aluminium oxide into aluminium and oxygen.

Hall and Héroult's process soon caused the price of aluminium to fall steeply, but at first there was little interest in the metal. The French plant using Héroult's patent used the aluminium only to make the copper alloy aluminium bronze. It was not until 1888 that the first aluminium-producing companies were established, one each in France, the USA and Switzerland.

The production of aluminium consists of five principal stages: the mining of the bauxite ore; the extraction of aluminium oxide from the bauxite; the manufacture of the carbon anodes for electrolysis; the reduction of aluminium oxide in electrolytic cells to form molten aluminium, to which may be added melted recycled aluminium scrap; followed by purification of the metal, and casting into ingots or the production of various alloys.

BAUXITE

Bauxite consists of aluminium oxide combined with water in varying proportions. The best-known composition is aluminium trihydrate, $Al_2O_3 . 3H_2O$, but substantial amounts of the monohydrate $Al_2O_3 . H_2O$ are available. The ore contains impurities in the form of silica (sand) and oxides of iron and titanium; the iron oxide gives it a red colour. Bauxite is named from the district of Les Baux, near Arles in the south of France, where it was first worked commercially.

Over 40 per cent of the world's bauxite is mined in Australia. Weipa in the north of Queensland, Gove in the Northern Territories and the Darling range of Western Australia are the major bauxite areas. Guinea in West Africa comes next, with 18 per cent, followed by Jamaica with 11 per cent. There are some small deposits in Brazil with nearly 100 per cent aluminium oxide, but the world's major reserves of high quality bauxite contain 40 to 60 per cent aluminium oxide – enough for the next 300 years. At a lower level of quality with 35–40 per cent aluminium oxide, the deposits will be sufficient for another thousand years, while clay and other minerals containing aluminium compounds will be at our disposal indefinitely.

FROM BAUXITE TO ALUMINA

Most bauxite deposits lie near the surface, in equatorial regions where sun and heavy rain have weathered the ore for millions of years. After vegetation and topsoil have been removed, the overburden of rock, sand or clay is stripped away, either by mechanical scrapers or powerful hydraulic jets; the bauxite is then mined by open-cast methods. Instead of removing the impurities by slagging during smelting, as in the production of iron, the aluminium ore has to be purified before it can be converted into metal. The concentration process devised by Karl Josef Bayer in 1888 depends on the fact that aluminium oxide dissolves in hot caustic soda (NaOH), but the impurities do not. This enables almost pure aluminium oxide to be separated.

The ore is crushed and ground to a fine powder, formed into a slurry with water, then mixed at high temperature with a solution of caustic soda in digesters, under pressure. The oxide dissolves, forming sodium aluminate which, being soluble, can be passed through filters, leaving the insoluble impurities behind. The aluminate solution is pumped into precipitator tanks about 10 metres in diameter and 20 metres high. Then very fine and pure particles of aluminium trihydrate are added as 'seed'. Under agitation by compressed air and with gradual cooling, pure aluminium trihydrate precipitates on the seed and is separated from the caustic soda solution by settling and filtering. It is then heated to about 1000°C, which drives off the chemically combined water, leaving aluminium oxide as a white powder suitable for electrolytic smelting. The caustic soda is recovered and returned to the start of the process, where it is used over again to treat fresh bauxite. Often the ore is converted to alumina near the mine and transported to the smelting plant, which is generally situated in a region where there is ample hydroelectric or other cheap forms of power.

MAKING THE ANODES

Two methods of anode manufacture are employed. Pre-baked anodes are formed by blending petroleum coke aggregate, pitch and scrap anodes at a temperature a little over the softening point of the pitch, and moulding it into blocks complete with pre-formed electrical connection sockets. The assemblies are fired at about 1100°C. A typical anode block is 70 cm wide, 125 cm long and 50 cm high. The other type is known as the Söderburg anode. Metal moulds containing carbonaceous material are mounted over the reduction cells, so that the contents are baked by the outgoing heat. Söderburg anodes require less energy to produce than pre-baked anodes, but they are less efficient in the reduction cells and so are becoming phased out in many aluminium smelting plants.

In most aluminium smelters, the carbon block anodes are joined to electrical connectors of steel or cast iron attached to aluminium. The anodes are lowered into the electrolytic cells to replace the anodes which have been consumed by the production of the aluminium. All plants get through a considerable number of anodes. For example, in Alcan's Northumberland plant about 110 000 anodes are consumed each year.

ALUMINIUM SMELTING

Reduction of alumina to aluminium takes place in steel container cells measuring 8–16 metres long, 4–5 metres wide and 1.5 metres high. They are lined with refractory bricks and a layer of carbon blocks which form the cathode. Carbon anodes – twenty in a typical plant – are suspended in each cell. These work in a series of between 150 and 200, called a pot line. Two lines of 200 cells would each produce about 170 000 tonnes a year. *Figure 6* illustrates a typical electrolytic cell.

Until the beginning of the Second World War, cryolite, mined in Greenland, was used in aluminium production, but since then synthetic cryolite has replaced the naturally occurring mineral. The cells contain molten cryolite to which some aluminium fluoride and calcium fluoride have been added to lower the melting point to about 940°C.

Alumina is fed through hoppers into the cells and dissolves in the molten cryolite. Direct current passes from anode to cathode. The current ranges from 200 000 to 300 000 A, depending on the plant's scale of operation. The amount of metal produced in each cell is proportional to the current. The passage of the strong electric current through the cell results in the evolution of heat in the same way as passing electricity through the elements of an electric fire generates heat.

The current breaks down the alumina into aluminium and oxygen, which reacts with the carbon to form carbon dioxide. The aluminium is attracted to the carbon cathode lining of the cells, and is syphoned into crucibles and transported to open-hearth furnaces to which reclaimed aluminium is added. Theoretically, 0.33 kg of carbon is required per kilogram of aluminium, but in practice the consumption is nearer to 0.4 kg per kilogram; the anodes must be replaced daily.

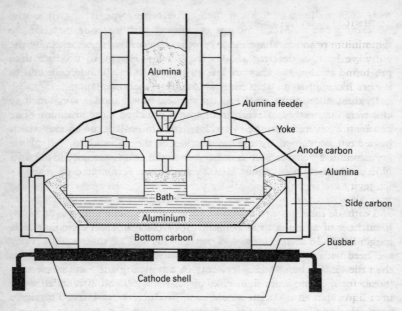

Fig. 6. Extracting aluminium by electrolysis (*Courtesy Elkem Aluminium ANS, Norway*)

The materials required to make aluminium illustrate that its smelting is a gigantic exercise in mass-handling. Thus 4½ tonnes of bauxite is needed to yield 2 tonnes of alumina by the Bayer process, which requires 1 tonne of fuel oil and uses 160 kg of caustic soda. In the electrolytic process about 14 500 kWh of electricity, 400 kg of carbon anodes and about 50 kg of cryolite are consumed per tonne. Each stage of production, from bauxite to aluminium, requires a great deal of energy, but the major part, about 65 per cent, is represented by the electricity used for smelting in the electrolytic cells.

The economic production of aluminium depends on the availability and cost of electrical energy. The USA, Canada, Norway, Brazil and Venezuela have ample hydroelectricity, which makes large-scale production possible. The UK, short of such power, depends for aluminium production on electricity from nuclear energy in Anglesey and a coal-fired station in Northumberland. In Australia the major source of energy comes from brown coal, but hydroelectricity is also used.

TECHNICAL DEVELOPMENTS

Aluminium producers have made continuous improvements in efficiency. Fifty years ago nearly 20 000 kWh of electric power was required per tonne of aluminium. Now the average is 14 500, though some producers have achieved 13 500 kWh. Computer-controlled pot lines, with individual microprocessors, have been installed. The handling, mixing and weighing of the materials from which the anodes are made are done automatically, with computerized control of the high temperature required to bake them. The anodes are changed automatically.

A great deal of research has gone into the improvement of pot life. The world average is about seven years. It has been found that a pot's longevity is determined mainly by the efficiency of its initial curing, but it is also affected by the way in which the lining is installed. Anode and cathode manufacture is being continually improved; in one development titanium diboride has been incorporated in cathodes to give them longer life.

There have been many efforts to find less energy-intensive methods than the Hall–Héroult process. Alcoa of America opened a pilot plant in Texas for making aluminium chloride, gasifying it, filtering it, allowing it to liquefy in an inert gas atmosphere, then electrolysing the liquid to produce aluminium. The process, which was described as a plumber's nightmare, consumed less electricity than Hall–Héroult and did not require cryolite, but eventually it was discontinued.

ALUMINIUM PRODUCTION AND THE ENVIRONMENT

In common with most other industries, aluminium producers have responded to environmental problems. Spent pot linings (known as SPL) from aluminium reduction cells contain 30 per cent carbon, 30 per cent refractory material and about 40 per cent cryolite, plus small amounts of impurities. In Australia about 30 000 tonnes of SPL is produced each year. Most is stored near the smelting plants or dumped in landfills, but in the future it will be necessary to treat SPL by combustion or some other process. In the USA, Congress initiated and is supporting 'Superfund Sites' for cleaning up polluted industrial areas. Four American smelters have benefited from this, and are developing improved methods of treatment. Among other environmental measures, the use of the Söderburg anode is gradually being phased out, in spite of the fact that it is cheaper to use than pre-baked anodes, because it emits harmful fumes. There are still several smelters, including some in the former USSR, which employ Söderburg anodes but the trend to look more to the environment continues.

One of the most important achievements of the industry has been the reclamation of aluminium by recycling soft-drink and beer cans, which has highlighted the vast energy savings that can be made. Tonne for tonne, it takes only a twentieth of the energy to recycle aluminium as it does to smelt it from the ore.

5

ALLOYS

An alloy is an intimate blend of one metal with one or more other metals or non-metals. For example, one type of brass contains two-thirds copper and one-third zinc; steel is iron alloyed with carbon. Several thousand alloy compositions are in use, kept under review by the International Standards Organisation and similar bodies. Documents specify the minimum and maximum limits of composition and the mechanical and physical properties required of each alloy.

An automobile contains many different alloys. Over half the weight of most cars consists of more than ten varieties of steel, including mild steel for bodywork, stainless steel for trim, and heat treated alloy steels for gears, connecting rods, camshafts and valves. Cast iron or aluminium alloys are used for cylinder blocks. The gearbox transmission housing, pistons and inlet manifold are of aluminium alloys. Some cars now contain aluminium alloy 'spaceframes' which form the structure around which the bodywork is assembled. The wheels may be of steel, aluminium alloy or magnesium alloy, and the wheel balance-weights are of lead alloy. There is a considerable amount of lead alloy in the battery. In some cars, especially Volkswagens, magnesium alloys are used for crankcases and transmission housings. The spark plugs of some cars have a nickel core surrounded by copper, but in expensive cars they have a nickel core surrounded with a platinum or gold–platinum alloy covering.

Most of the metallic materials in cars are alloys. The only pure metals are copper for electrical wiring and some washers and gaskets; a minute amount of nickel appears in the electroplating and in some spark plugs. Although pure metals have useful properties such as high conductivity of heat and electricity, they are not used in structural or mechanical engineering because their strength is insufficient for the arduous duties which materials are expected to perform.

HOW ALLOYS ARE MADE

Some alloys are made direct by smelting; thus ferro-chromium and ferro-manganese, which are required in large quantities by the steel industry, are made by smelting mixed ores. However, most alloys are prepared by mixing two or more metals in the molten state; they dissolve in each other and the alloy so formed is poured into ingot moulds and allowed to solidify, or transported molten to foundries where it is made into castings. Generally the major ingredient is melted first, then the others are added to it. A plumber making solder might melt lead, add tin, and cast the alloy into sticks or bars. Some pairs of metals do not behave in this way: if molten aluminium and lead are put together they behave like oil and water; when cast the metals separate into two layers.

One difficulty in making alloys is that metals have different melting points. Thus copper melts at 1083°C, while zinc melts at 419°C and boils at 907°C; so, in making brass, if we were simply to put pieces of copper and zinc in a crucible and heat them together above 1083°C, both metals would certainly melt, but at that high temperature the liquid zinc would boil and the vapour would oxidize in the air. To overcome this difficulty the metal with the higher melting point, copper, is heated first. When this is molten, solid zinc is added and is quickly dissolved in the liquid copper before very much zinc has vaporized. Even so, allowance has to be made for unavoidable zinc loss – about one part in twenty.

Sometimes making alloys is complicated because the metal with the higher melting point is in the smaller proportion. For example, one light alloy contains 95.5 per cent aluminium (melting point 659°C) with 4.5 per cent copper (melting point 1083°C). If the small amount of copper were melted first it would have to be so overheated to persuade twenty times its weight of aluminium to dissolve in it that gases would be absorbed, leading to unsoundness. In this, as in many other cases, the alloying is done in two stages. First an intermediate 'hardener alloy' is made, containing 50 per cent copper and 50 per cent aluminium; this alloy has a melting point considerably lower than that of copper and, in fact, below that of aluminium. Then the aluminium is melted and the correct amount of hardener alloy added; thus, to make 100 kg of the aluminium–copper alloy, 91 kg of aluminium would be melted first and 9 kg of hardener alloy added to it.

Iron from the blast furnace is converted to steel in oxygen furnaces, and the steel is then transferred to a ladle so that ferro-manganese and other ferro-alloys can be added to obtain the required composition. In making one type of stainless steel, ferro-chromium containing 52 per cent chromium, 7 per cent carbon and the remainder iron is added.

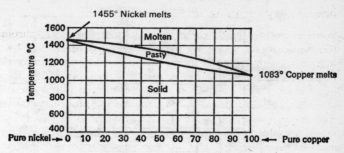

Fig. 7. The melting points of the copper–nickel alloys

When alloys are being made, a working margin of composition is allowed for the major constituents and the minor impurities, so that the alloy will have the required properties. Each alloy system is considered on its merits. For example, some aluminium alloys which are to be cast are allowed to have from 0.5 to 1.5 per cent of iron. On the other hand, the iron content of zinc diecasting alloys is limited to less than 0.1 per cent.

THE MELTING POINT OF ALLOYS

In winter, when ice forms on the road, salt is spread to melt the ice. This practical use of a natural phenomenon demonstrates that a mixture of salt and water has a lower freezing point than that of pure water. A similar effect is to be found in metallurgy, for when one metal is alloyed with another, the melting point is always affected.

The melting point of some alloys can be worked out approximately by arithmetic. For instance, if copper (melting point 1083°C) is alloyed with nickel (melting point 1455°C) a fifty–fifty alloy will melt at about halfway between the two temperatures. However, such an alloy does not melt or freeze at one fixed and definite temperature, but progressively solidifies over a range of temperature. Thus, if a fifty–fifty copper–nickel alloy is melted and then gradually cooled, it starts freezing at 1312°C; as the temperature falls, more and more of the alloy becomes solid until finally, at 1248°C, it has solidified completely (*Fig. 7*). This 'freezing range' occurs in most alloys, but it is not found in pure metals or in some special alloy compositions known as eutectics, which melt and freeze at fixed temperatures.

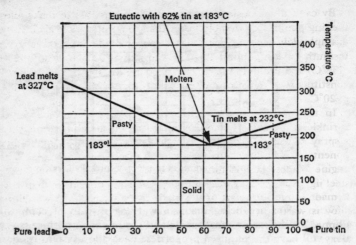

Fig. 8. The melting points of the tin–lead alloys

Alloys of tin and lead provide an example of one of these special cases. Lead melts at 327°C and tin at 232°C. If lead is added to molten tin and the alloy is then cooled, its freezing point is lower than the freezing points of both lead and tin (*Fig. 8*). When a molten alloy containing 90 per cent tin and 10 per cent lead is cooled, it reaches a temperature of 217°C before beginning to solidify. Then, as the alloy cools further, it gradually changes from a completely fluid condition, through a stage when it is like porridge, and finally, at a temperature as low as 183°C, the whole alloy becomes completely solid. By referring to *Fig. 8* it can be seen that, with 80 per cent tin, the alloy starts to solidify at 203°C and finishes only when the temperature has fallen to 183°C (note the recurrence of the 183°C).

At the other end of the series an alloy with 20 per cent tin, a typical solder, starts to freeze at 270°C and is completely solid at the now familiar temperature of 183°C. One particular alloy, containing 62 per cent tin and 38 per cent lead, melts and solidifies entirely at 183°C, without any freezing range. Similar effects occur in many other alloy systems; the special composition in each series which has the lowest freezing point and which entirely freezes at that temperature is known as the eutectic alloy. The word is derived from the Greek *eutektos*, 'easily melted'.

By carefully choosing constituents, it is possible to make alloys with melting points considerably lower than those of the metals involved. For example, low melting point alloys can be made from these five metals: bismuth (melting point 261°C), cadmium (321°C), lead (327°C), tin (232°C) and indium (156°C). By selecting different proportions of these constituents, a range of melting points can be selected, from 300 down to 20°C.

In the past such low melting point alloys were used for anti-fire sprinklers; the heat of a conflagration caused the alloy to melt, releasing a spray of water. Nowadays a far wider application is for holding components of complex shape for accurate machining processes. When aero engine blades are machined it would be difficult to hold them in a steel jig, as is used for machining components of regular shape. A mould is made around the part of the blade to be secured: a low melting point alloy is warmed, and is melted around the component. This mould is gripped during machining, after which the alloy is warmed and melts away. Other low melting point alloys, mainly based on bismuth, enable bending and forming operations in tubes to be done without them flattening or buckling.

THE STRENGTH OF ALLOYS

If sand is added to sugar the mixture displays the properties of both ingredients: it is sweet but gritty, and the more sand in the sugar the less pleasant the taste. Metals rarely behave in such a predictable manner. When one metal is added to another, the alloy has a new individuality which one might not expect from the properties of the two metals. A weak or soft metal alloyed with another element may produce a strong alloy with strikingly different properties from those of the parent metal. For example, pure iron is a comparatively soft metal, and carbon in its commonest form is weak, but as little as 0.5 per cent carbon in iron produces a strong steel which responds to hardening and tempering treatments, and can be used for making highly stressed components of automobiles, aircraft and a multitude of other machines.

The addition of 5 per cent aluminium to copper produces an alloy twice as strong. With 10 per cent, the aluminium bronze produced is three times as strong. Naturally, the temptation would be to go on adding more and more aluminium in the hope of getting still more spectacular increases of strength. Unfortunately, this does not occur. With 10 per cent aluminium, the aluminium bronze is as strong as mild steel – very tough, and in many ways a valuable engineering alloy.

At 16 per cent aluminium, the alloy is about as brittle as a carrot and much less useful. There is a no-man's land between 16 and 88 per cent aluminium where the alloys are valueless for engineering purposes. (The fifty–fifty hardener alloy mentioned on page 38 is certainly used for making up alloys, where its relatively low melting point and its brittleness are convenient. It is, however, only an intermediate product, not an engineering material.) At the other end of this copper–aluminium series, the alloys containing up to 12 per cent copper are stronger than pure aluminium, and they too are used in industry.

In addition to the increase of strength achieved by alloying, many alloys respond to heat treatment by developing still better mechanical properties. Whenever some special property is required – great strength, toughness, resistance to wear, high electrical resistance, magnetic properties or corrosion resistance – it is usually possible to produce it in a carefully chosen, skilfully made and intelligently manipulated alloy.

SUPERPLASTIC ALLOYS

About sixty years ago it was discovered that some alloys show unusually high ductility at raised temperatures. Later the term 'superplasticity' was coined by Russian metallurgists. At first the phenomenon was thought to be of little practical use, but in 1962 a zinc alloy containing 22 per cent aluminium was heated to 300°C, quenched to room temperature and reheated to 250°C. The grain structure of the alloy gave it the property that under stress it could be stretched as much as ten times without cracking. Sheets of the material could be heated and pressed easily into complex shapes, using press tools much simpler and cheaper than those normally required.

After that discovery experiments were conducted to find other alloys that exhibited the same effect. Now the most widely used superplastic alloy is based on aluminium with 6 per cent copper and 0.4 per cent zirconium. Sheets of this material can be clad with pure aluminium. One of the new aluminium–lithium alloys, which will be discussed on page 270, is superplastic and also heat-treatable.

These materials are so plastic when heated that they can be pressed into moulds by air pressure. A sheet, preheated to about 500°C, is clamped in position in the mould and pressed to the required form. With low tooling costs, superplastic alloys are ideal for medium production runs; complex shapes formerly fabricated from several pieces can be produced in one operation. A recent development has been the production of the engine intake leading edge for the British Aerospace Hawk trainer/combat aircraft. Previously this was built from eight components, but when the superplastic alloy was used it was made in one piece. About half of superplastic aluminium alloy production is for aircraft, but the alloys are also now being used for medical equipment, architectural panels, specialist car bodies and communications dishes. Recently, railway carriages have been reduced in weight by the use of superplastically formed aluminium alloys.

6

METALS UNDER THE MICROSCOPE

The microscope is the metallurgist's most useful scientific instrument. With its aid a skilled observer can learn about the structure, the manufacturing history, the effect of heat treatment and the cause of fracture of a piece of metal or alloy. The optical microscope in a works laboratory is capable of magnifying up to about 1500 times, and by the use of different objective lenses a metal can be examined under several magnifications. Usually a camera is provided so that the structure of the metal can be photographed. Modern research laboratories also have several different types of electron microscope with vastly greater resolution, and these will be described at the end of the chapter.

As metals are practically opaque, the optical metallurgical microscope is designed to examine their surfaces by reflected light. A specimen of convenient size, about $10 \, mm^3$, is cut off. One face is filed flat and then ground with successively finer grades of carborundum papers which are lubricated with paraffin or water and supported on plain glass surfaces. Final polishing for optical viewing is continued on cloth impregnated with a slurry of fine alumina or magnesia, to produce a mirror-like finish on the metal. Recently several refinements have been introduced in the preparation of specimens. Silicon carbide, boron nitride and diamond dust are being used as the abrasives, which helps to give better reflective surfaces than before. The cloths used for polishing are selected from a range of textile materials to suit the hardness of the specimen.

First the polished surface is examined under the microscope to see if there are any cracks or inclusions. *Figure 9* shows the appearance of a piece of polished wrought iron such as was made in the nineteenth century for chains, horseshoes or the gates of ancestral homes. The dark streaks are inclusions of slag which was present in this metal due to its method of manufacture, described on page 114.

If a metal has failed in service, the position and shape of the crack can be examined and the cause of the failure diagnosed. For example, it may

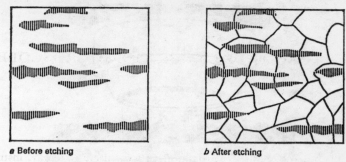

a Before etching   *b* After etching

Fig. 9. The microstructure of wrought iron

be found that the crack began at a part of the metal object that had been welded inexpertly. Such an examination, useful though it is, does not indicate the structure of the grains in a metal component, and it tells little of the success or otherwise of heat treatment. Much more information can be obtained by etching the polished surface of the metal and examining it under the microscope. The specimen is immersed in a chemical solution, chosen according to the metal and the particular constituent it is desired to observe.

For general examination, steel is usually etched in a mixture of 98 parts alcohol and 2 parts nitric acid, though if one constituent, known as 'cementite' (page 296), has to be identified, the steel is etched in boiling sodium picrate solution, which makes the cementite appear black. The time of etching varies from a few seconds to half an hour, depending on the alloy and on the solution employed. Sometimes more sophisticated methods are used, such as electrolytic attack or distilling away some of the metallic surface in a vacuum or by slightly heating the specimen to form thin films of oxide.

THE GRAIN STRUCTURE OF METALS

Microscopic examination of the etched surface of a polished metal reveals that it is built up of innumerable small 'grains'. *Figure 9b* shows the same piece of wrought iron as in *Fig. 9a*, but this time the metal has been etched in a dilute solution of nitric acid in alcohol. (The illustration represents a section cut through the metal; in reality the grains are three-dimensional.)

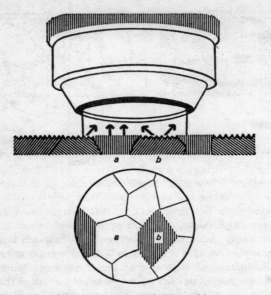

Fig. 10. Grain structure revealed by etching

Like many other technical words, 'grain' has a somewhat different meaning from its everyday use, which connects grains with sand, salt or sugar. A grain of metal differs from one of sugar in at least two respects. First, each grain of sugar is brittle and can be crushed into powder, whereas metal grains are usually ductile. Second, there is little cohesion between the grains of a lump of sugar, while in a block of metal great force is needed to separate the grains. Sometimes this cohesion of the boundaries may be weakened; for example, when one part in ten thousand of bismuth is present in copper or gold, the bismuth distributes itself at the grain boundaries, thus reducing cohesion and causing the piece of copper or gold to be brittle.

On etching the metal, the grain boundaries are more readily attacked than the interior of the grains. Tiny channels are eaten away at the extremities of each grain; when the etched metal is examined under the microscope, light falling on these channels is scattered, so that the boundaries around the grains appear dark. Under some conditions of etching, whole grains appear contrasted in tone, which may be even more distinguishing than the demarcation of the boundaries. *Figure 10*

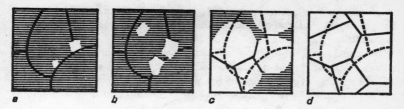

Fig. 11. Successive stages of recrystallization

illustrates the cause of this contrast. It will be seen that the grains have been attacked by the etching solution in different ways. When light falls on the grain marked '*a*' most of it will be reflected back through the objective of the microscope and consequently this grain appears bright. On the other hand, light falling on grain '*b*' is reflected sideways, and so the whole of it appears darker.

Although the grain structure of most metals is revealed only after etching and with the aid of the microscope, there are a few examples where large metal grains can be seen with the unaided eye. When brass articles such as door knobs, or the long brass handles at the entrances to public buildings, have been in use for a year or so, the constant rubbing with human hands polishes and etches the brass, so that the grains are revealed as small patches or spangles of slightly varying colour tones. In such brass knobs and handles the grain size is comparatively large, of the order of two millimetres. Large grains of zinc can also be seen on galvanized steel articles.

However, in most metals and alloys the grain structure can be distinguished only by the use of a microscope. For example, some steels, in the condition in which they are used in industry, have grains about 0.025 mm across. The grain size of a metal depends on the casting temperature, the impurities present, and the mechanical working and heat treatment to which the metal has been subjected; this is one of the important ways in which metallurgists can fit metals for the tasks they will have to perform in service. A metal with fine grains will be somewhat harder and stronger than one with coarse grains.

After a metal has been 'cold worked' – drawn into wire or rolled into sheet – it is possible to bring about the birth of new grains by heating the metal. This is illustrated in *Fig. 11* and the process is known as 'recrystallization'. The first effect of heating is to form small, new grains as shown in white in *a*, and these rapidly enlarge until further growth is restricted by one new grain meeting another, as shown in *b* and *c*. Ultimately the original system of grains is obliterated and the new, recrystallized structure is shown in *d*, the original grains being indicated in the drawing by dotted lines.

On continuing the heating, adjustments of the boundaries may take place, resulting in the further growth of some grains at the expense of others. The eventual size of the grains in a piece of metal depends on the amount of deformation previously existing and on the time and temperature of the heating process.

SINGLE CRYSTALS

By special means it is possible to solidify molten metal in such a way that it forms one large grain, known as a single crystal. Over fifty years ago small single crystals had been produced in low melting point alloys of bismuth, but it was not until the late 1970s that the technology advanced dramatically. Now large components made of single crystals in high strength alloys can be made in considerable quantity. They have the great advantage of superior mechanical properties because they have no grain boundaries. In the production of silicon chips, single crystals 2 m long and 20 cm in diameter are produced, and then sliced into wafers for making the chips.

A common method for the production of large single crystals is first to select one grain of the metal. Suitable grains are examined by X-rays until one is found with the required crystal orientation, with a minimum of discontinuities and as near perfect as possible. This selected grain, known as the seed, is attached to a holder which is then positioned above the container of molten metal.

The temperature of the molten metal is adjusted so that the centre of the upper face of the liquid is exactly at its freezing point. Then the seed is carefully lowered until one face touches the metal. This attaches itself to the seed, which is slowly rotated and very gradually withdrawn – only a few centimetres per hour. In this way the original seed eventually becomes one large single crystal. In the production of silicon chips the operation is done in a controlled atmosphere; with engineering alloys the operation is done in vacuum.

From this very basic description one can imagine how a single crystal block can be made as the first stage in the manufacture of silicon chip, but in recent years the technology has been extended to produce complex single crystal components for jet aircraft. Usually the materials are superalloys; a typical composition is nickel-based with 5 per cent cobalt, 10 per cent chromium, 4 per cent tungsten, 5 per cent aluminium, 12 per

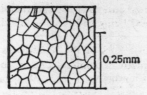
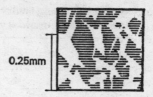

Fig. 12. The microstructure of a copper–nickel alloy

Fig. 13. The microstructure of 60/40 brass

cent tantalum and 1.5 per cent titanium. *Plate 5a* shows a segment of a turbine blade for a fighter aircraft engine. In February 1992, the company producing this part announced that it was their millionth single crystal casting.

Plate 5b shows two cutaways of turbine blades. The one on the left was produced as a single crystal casting. The component on the right was of polycrystalline structure and was produced by directional solidification controlled so that the grains grow continuously from one end of the casting to the other, in exactly the required alignment.

Single crystal technology has made it possible to operate jet aircraft engines at increased turbine inlet temperatures, achieving improved fuel economy and increased power. The technology is now also being used for land-based turbine blades, some about 70 cm long.

THE STRUCTURE OF ALLOYS: SOLID SOLUTIONS

The properties of any metal are altered when it is alloyed with another, and this makes one wonder whether microscopic examination will reveal structural differences between pure metals and alloys. In some cases such differences can be observed; for example, two constituents can usually be identified in lead–tin or iron–carbon alloys. But other alloys may consist of the polyhedral grains characteristic of a pure metal, and only one constituent is discernible. For example, the microstructure of a copper–nickel alloy is shown in *Fig. 12*. If this alloy were composed of grains of copper mixed among grains of nickel, it would be possible to distinguish them by colour alone. Clearly, then, something must have happened to the copper and nickel atoms to mingle them so closely that microscopic examination cannot reveal the individual metals.

The two metals are said to exist in a state of 'solid solution'. It may seem strange that one solid metal can exist in solution in another, but there is a wider definition to the word 'solution' than merely 'something

Table 4. The solid solubility of various metals in magnesium
at room temperature and at 300°C

Element	Solid solubility in magnesium (weight per cent)	
	Room temperature	300°C
Aluminium	2.3	5.3
Calcium	about 0.1	0.18
Copper	under 0.1	0.1
Lead	3.7	16.0
Manganese	0	0.1
Silver	1.5	3.6
Zinc	1.7	6.0

dissolved in a liquid'. A solution may be described as an intermingling of
one substance in another so closely that the dissolved substance cannot
be distinguished or separated by mechanical means. This description can
be applied to the condition of the solid copper–nickel alloy, just as it is
applied to salt dissolved in water.

Only a few pairs of metals, such as copper and nickel, can exist in solid
solution throughout the whole range of possible compositions, but most
metals can contain at least some of another metal in solid solution. In an
alloy where the metals do not show complete solid solubility, separate
constituents may be recognized by examining the etched alloy under the
microscope. *Figure 13* shows the microstructure of a brass containing
about 40 per cent zinc, similar to that used for domestic water taps or
brass nuts. The two constituents are two solid solutions, each of different
composition.

Just as tea or coffee dissolve more sugar when hot than when cold, so
most metals retain more of another metal in solid solution when they are
hot. *Table 4* gives the solid solubility of various metals in magnesium at
room temperature and at 300°C. The solid solubility of each metal is
greater at the higher temperature.

INTERMETALLIC COMPOUNDS

In aluminium–copper alloys, solid aluminium at about 530°C can retain
5 per cent copper in solution, while at room temperature it can normally
hold less than 0.5 per cent. Therefore, if an alloy containing, say,

4 per cent copper is slowly cooled from about 530°C, it comes to a stage, at just below 500°C, when it can no longer hold that amount of copper and, as the temperature falls further, less and less copper can be retained in solution. The surplus does not separate out as distinct grains of that metal, but in the form of an intermetallic compound to which the symbol $CuAl_2$ is given.*

As the temperature becomes progressively lower, more and more copper comes out of solution and forms $CuAl_2$, so that finally the slowly cooled alloy at room temperature consists of a background of aluminium containing only a small amount of copper in solid solution, together with a number of particles of the intermetallic compound $CuAl_2$ dispersed throughout the alloy. When isolated, this compound is found to have characteristic properties: it is, for example, extremely hard.

When the composition of an alloy is such that its structure consists of a matrix of solid solution and particles or grains of an intermetallic compound, the alloy is likely to to be useful, for it combines the toughness of the solid solution with the hardness of the intermetallic compound. *Plate 31a* shows the microstructure of a lead–tin–antimony alloy that is sometimes used for bearings. The 'cubes' are an intermetallic compound of tin and antimony.

In recent years there have been important discoveries in the field of intermetallic compounds. The nickel–aluminium compound Ni_3Al displays the unusual property that up to 700°C it increases in strength with increasing temperature. In early researches this compound was found to be brittle, which limited its applications, but Japanese metallurgists doped the Ni_3Al with 500 parts per million of boron, and even more minute amounts of chromium, hafnium and zirconium, which made the compound much more 'workable' so that it could be used, for example, in making dies for compressing the powders from which permanent magnets are made. Two other compounds, each of samarium and cobalt ($SmCo_5$ and Sm_2Co_{17}), are used for the small magnets in electronic wristwatches.

EQUILIBRIUM CONDITIONS

The behaviour of the aluminium–copper alloys will be referred to again on page 157 when age hardening is discussed, but it may be remarked

---

* $CuAl_2$: its composition by weight is 54 per cent copper, 46 per cent aluminium. The symbol denotes that one copper atom is intimately associated with the two aluminium ones.

here that the constituents completely separate only on slow cooling. If the aluminium alloy containing 4 per cent copper is quenched in water from 500°C so that it is rapidly cooled to room temperature, the copper does not have time to come out of solid solution and the aluminium at room temperature is forced to hold a surplus of copper in solid solution. In other words it is 'supersaturated', and it may be some days after the quench before the copper atoms spontaneously separate. In some other alloys treated in this way a state can be reached after quenching where adjustment can be attained only by warming the alloy, which gives opportunity for the separation to take place.

Many of the phenomena of metallurgy may be attributed to the sluggishness of alloys in attaining equilibrium. The hardening and subsequent tempering of steel and the age hardening of aluminium alloys depend on the fact that rapid quenching produces a different condition from that produced by leisurely cooling in a furnace.

THE STRUCTURE OF EUTECTICS

There is another type of constituent seen under the microscope which may appear in those series of alloys which form eutectics. We saw on page 40 that in certain alloys there is a particular composition which melts at a lower temperature than any other alloy in that series. This composition is known as the eutectic. The structure of eutectics may consist of alternate thin layers of the metals concerned, or in other cases small globules of one metal embedded in a matrix of the other. For example, the silver–copper eutectic is composed of 72 per cent silver and 28 per cent copper. If an alloy in such a series is not of eutectic composition, the structure seen under the microscope consists of one of the metals and some eutectic. Thus an alloy of 10 per cent silver and 90 per cent copper has a structure of grains of copper plus a small amount of eutectic. Similarly, at the other end of the series, an alloy with 90 per cent silver and 10 per cent copper consists of eutectic plus silver. The nearer the composition approaches that of the eutectic, the greater the proportion of eutectic seen when the alloy is viewed under the microscope. *Plate 31b* shows the appearance of the alloy containing 60 per cent copper and 40 per cent silver.

THE BIRTH OF GRAINS FROM MOLTEN METALS

Most people have seen the attractive patterns formed in winter when water vapour freezes on windows. A similar structure may be formed

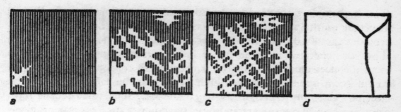

Fig. 14. Successive stages in the freezing of a molten alloy, showing the formation of dendrites

when a liquid alloy solidifies, though the alloy has to be polished, etched and examined under the microscope to make its structure apparent. When a molten alloy begins to freeze, minute crystals form at various points in the liquid and these start to grow by developing branches in certain directions, as shown in *Figs. 14a* and *b*. These tree-like formations are known as dendrites. When the arm of a dendrite meets that of another, as in *c*, outward growth is restricted, but the spaces between the branches continue to fill in until all the metal is solid, as in *d*.

When a solid piece of *pure metal* is sectioned, polished, etched and examined under the microscope, no dendrites can be seen because the metal is uniform. But when a cast *alloy* which has solidified is examined microscopically, evidence of dendrites can usually be observed, for the composition of the first part of the alloy to freeze differs from that which finally freezes, and what is known as a 'cored structure' is produced. If such an alloy is then worked either hot or cold and is heat treated, this cored structure may be gradually eliminated and a homogeneous solid solution formed having the uniform structure shown in *Fig. 12*.

MODERN METALLOGRAPHIC INSTRUMENTS

Fifty years ago a metallurgical microscope that could give a magnification of 1500 was considered powerful; then it became possible to photograph the microstructures of metals at magnifications of 7000 by the use of ultraviolet light. Next microscopes using beams of electrons were developed, and soon magnifications of over 100 000 became possible. The photograph of dislocations in cold-worked aluminium in *Plate 32a* was taken at a magnification of 50 000.

The latest electron microscopes can magnify more than a million times, and can resolve details in very thin specimens down to two ten-millionths of a millimetre, on which scale individual atoms are visible.

Plate 6 shows a transmission electron microscope (known as a TEM). The tank at the top left of the picture is the voltage generator which provides the 400 000 volts to accelerate the electron beam. The column above the operator contains the series of electromagnetic 'lenses' that focus the electron source onto the sample and then magnify the image on a fluorescent screen. There is a high vacuum in the column because the electrons can only travel in those conditions. Such microscopes have greatly improved our understanding of phenomena such as crack propagation, defects in crystal structures, and the effects of impurities and of nuclear radiation on metals. Preparing the samples of the materials for examination is a skilled task, sometimes taking several days, because the sections have to be less than one ten-thousandth of a millimetre thick.

A different type of instrument, the scanning electron microscope (SEM), has the advantage that it does not require the fantastically thin specimens that must be produced for the transmission electron microscope described above. The SEM provides magnifications of about 100 000 and has a large depth of field to 'see' inside deep cracks. Many research activities have benefited from the use of the SEM in studies of crystal growth, corrosion, electroplated deposits and bonding processes. The SEM is now found in many industrial and research laboratories. Scanning electron microscopes are now often combined with X-ray microanalysis units which use the characteristic X-rays produced in each metal by an electron beam to give a very rapid chemical analysis of the material over an area only a micrometre square (0.001 × 0.001 mm). Such equipment is now often combined with computer control systems.

A still more sophisticated instrument, the scanning transmission electron microscope (STEM), combines the benefits of the two microscopes described above. These instruments can examine areas of a metal containing less than forty atoms, and some sophisticated ones can examine and analyse groups of only two or three atoms.

Another remarkable instrument, known as the scanning tunnelling microscope, was designed for the study of surfaces and is sensitive enough to produce images of individual atoms. A sharp tungsten point is brought to within one two-millionths of a millimetre from the surface to be studied. An electric current begins to flow before contact is made, and this current depends strongly on the distance from the point to the surface. The point is programmed to move in such a way that the current remains constant, and thus it 'draws' the positions of individual atoms on the surface. The changes in the position of the point map out minute areas of the surface, and the image of the atoms is projected on a monitoring screen from which photographs can be taken. This instrument is used to examine surfaces of a wide range of metals and semi-conductors. With its aid, it has been possible to select individual atoms, take them out of position and place them in adjacent positions.

In recent years the precise study of the surfaces of materials has become an important branch of science. Another device which has contributed to these studies is the scanning Auger microscope, illustrated in *Plate 7*. A beam of high-energy electrons is fired at the material to be examined, causing low-energy electrons to be ejected from the first few atomic layers. The energy of these electrons is very specific to the atoms from which they originated, and thus every element has its own 'Auger trademark'. At first the Auger instruments were used to monitor the purity of semiconductors and of surfaces prior to the deposition of coatings in individual processes, such as electroplating and the galvanizing of steel. More recently, however, scanning Auger microscopes have proved invaluable in the search for new superconductors, described on page 294. By the use of such methods, scientists are approaching the ultimate regions in which many basic phenomena of the science of metals may be studied.

7

THE INNER STRUCTURE OF METALS

All chemical elements, including the metals, are composed of atoms, which are the smallest particles retaining the individual characteristics of the element. The atoms are so small* that they cannot be seen, except by the most powerful electron microscopes. Using such instruments, physicists have been able to widen our knowledge about atoms and this has helped to explain the behaviour and properties of metals. It is important to know how the atoms are arranged in a grain of metal. Are they all piled at random, or are they arranged in a regular and orderly formation? Are the atoms of all metals exactly the same size, or does an atom of, say, lead occupy less space than an atom of magnesium?

About a hundred years ago scientists began to get evidence that the atoms of metals were arranged in regular patterns and belonged to the class of materials which are crystalline. This was proved in 1911, when Max von Laue examined metals with X-rays. Many years later it became possible to solidify molten metals at an almost incredible speed, so that when solid they were not crystalline but amorphous; this development is discussed on page 273. The arrangement of metal atoms in the crystalline condition has been compared to the marshalling of a regiment on a parade ground, and that in the amorphous condition to people bustling to and fro near a crowded railway station.

The concept that grains of metals are crystalline is sometimes a stumbling block to those who are beginning to study metallurgy. Most people associate crystals with substances, such as diamond, quartz and copper sulphate, which are hard and sparkling, but this does not imply that all crystalline materials have these characteristics. The basic defini-

---

* It has been calculated that a cubic millimetre of copper contains about 84 693 000 000 000 000 000 atoms, about 15 thousand million for every person in the world.

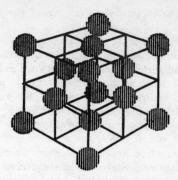

Fig. 15. Face centred cubic lattice

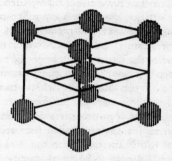

Fig. 16. Body centred cubic lattice

tion of crystallinity does not concern the outward appearance of a material but arises from its inner symmetry. Hard sparkling substances such as glasses are not truly crystalline, while lead, for example, which is far from sparkling, has a crystalline structure.

Metallurgists and physicists have discovered the different 'lattice' structures which atoms adopt in the various metals, the distance between neighbouring atoms, and the space the individual atoms occupy. In aluminium, copper, nickel, lead, silver, gold, platinum and several other metals, the atoms are spaced evenly in rows at right angles to one another. This can be likened to atoms arranged at each corner of millions of adjoining cubes, while other atoms occupy positions at the centre of each of the cube faces. This particular atomic lattice pattern is known as 'face centred cubic', and is illustrated in *Fig. 15.* The lines connecting the atoms in drawings like this are intended to emphasize the relative positions of the atoms.

In iron at room temperature, and in several other metals such as vanadium, tungsten, molybdenum and sodium, the atoms are arranged in another pattern. There is again an atom at each corner of each imaginary cube, but instead of other atoms lying at the middle of the cube faces, a single atom is located at the centre of every cube. This structure is known as 'body centred cubic', and is illustrated in *Fig. 16*. Iron is particularly interesting, for at room temperature its atoms are arranged in the body centred cubic form, but at 906°C the atoms reshuffle into the face centred cubic pattern, while at a still higher temperature, about 1400°C, the iron atoms change back to a body centred cubic lattice.

In zinc, magnesium, titanium and cadmium the atoms are arranged in hexagonal patterns. Most common metals have either face centred cubic, body centred cubic or hexagonal lattices, but some have more complex arrangements; for example, tin and samarium are tetragonal.

After having discovered how the atoms of pure metals were arranged, investigators turned their attention to the lattice patterns of alloys. From microscopic examination they already knew that the grain structures of alloys were different from those of the metals of which they were composed, so it seemed likely that some parallel difference might be found when the atomic lattices of alloys were investigated. There were interesting questions to be answered: for example, what happens to the atomic arrangement when a metal which crystallizes in the face centred cubic pattern is alloyed with one of a hexagonal type? Can any light be thrown on the cause of the increase of strength which occurs when one metal is alloyed with another? As an example, we shall look at the behaviour of the copper–zinc alloy, brass.

THE ATOMIC LATTICE PATTERNS OF BRASS

The atoms in pure zinc are in hexagonal arrangement whereas copper atoms form a face centred cubic lattice structure. The spaces occupied by the zinc and copper atoms differ, that of zinc being about 13 per cent larger than that of the copper atoms. When only small amounts of the zinc are present in brass, the prevailing atomic arrangement is similar to that of copper, which means that zinc atoms, in solid solution, have to adapt themselves to the face centred cubic pattern. Each zinc atom takes the place of one copper atom, but because the space occupied by the zinc atom is greater than that of copper, the face centred cubic lattice is

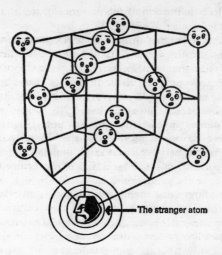

The stranger atom

Fig. 17. Distortion of a lattice by a different-sized 'stranger atom'

distorted at the point where the 'stranger atom' is introduced. A comparison between *Figs. 17* and *15* will demonstrate this better than words.

When progressively increasing amounts of zinc are alloyed with copper, the brass becomes increasingly hard; this can be explained by referring to the drawing of the 'stranger atom'. The introduction of a new atom of different size from the rest has brought about a distortion in the lattice, leading to a greater resistance to deformation than before. Such strengthening by alloying can occur whether the stranger atoms are larger or smaller than those of the original metal, for distortion of the lattice occurs in either case.

As more zinc is added, the face centred cubic arrangement of the atoms becomes increasingly distorted until, when about 36 per cent zinc is present, the lattice becomes unstable and here and there another form comes into existence, which is body centred cubic. Up to this point the strength and hardness of the brass gradually increase with rising zinc content, but at the 36 per cent composition the properties are sharply altered and further additions of zinc bring about a more rapid rise of hardness than before.

A change can also be noticed when the alloy is examined under the microscope. Up to 36 per cent zinc, the brass is in the form of a solid solution in which no direct indication of the presence of the zinc can be observed except by some change of colour. With over 36 per cent zinc a new constituent or 'phase' begins to appear, and can be seen under the microscope (see *Fig. 13* on page 49). It is necessary to distinguish between these phases, and they are given Greek letters to designate them. According to the usual practice, the first phase is called alpha (α), the next beta (β); a brass containing 36 to 42 per cent zinc includes both alpha and beta phases, and is called an alpha/beta brass. When still more zinc is added, further phases appear with different atomic lattices; and with each new structure the mechanical properties of the brass alter sharply. In all five different constituents can exist, though not more than two at once, in the copper–zinc alloys. The effects of increasing amounts of zinc alloyed with copper are summarized in *Table 5*, which compares the composition of the alloys with their microstructure, atomic lattices and mechanical properties.

The copper–zinc alloys are rather complicated because so many atomic reshufflings occur. Some alloy systems, such as the copper–tin series, are even more complex; others, such as the copper–nickel alloys, are straightforward because both metals have face centred cubic atomic lattices, their atoms are of similar size, and they are able to form a continuous range of solid solutions, the lattice dimensions changing gradually, from pure copper to pure nickel.

IMPERFECTIONS IN METAL CRYSTALS

From the previous description of the crystalline form of metals, it might be concluded that their atoms are arranged in perfect symmetry. However, experimental work during the last fifty years has confirmed that the crystalline formation in each metal grain is far from perfect, and that many of their valuable mechanical properties are produced by imperfections. In particular, the evidence explains why metals are tough and withstand heavy stresses, extending a little and then showing resistance to further deformation. It also helps to explain why metals will endure shock loads and stress reversals.

Such discontinuities are described as dislocations, but there are several associated effects, all of which have a profound influence on the strength of metals:

1 line defects, called dislocations, of which there are several million in each cubic millimetre of metal;

2 point defects, caused by atoms which are small enough to squeeze into the interstices between larger atoms;

Table 5. Some changes produced by alloying zinc with copper

Composition of alloy	Atomic structure		Crystal structure as seen under microscope	Zinc (per cent)	Mechanical properties *			
	Description of lattice	Spacing of atoms (angstrom units)			Diamond pyramid hardness number	Tensile strength (tons per in²)	(newtons per mm²)	Ductility (expressed as per cent elongation)
Pure copper	Face centred cubic	3.615	Grains of pure copper		53	15	230	45
Copper alloyed with up to 36% zinc	Face centred cubic pattern which progressively increases in size	3.615 with 0% zinc, increasing to 3.698 with 36% zinc	Grains of solid solution of zinc in copper (known as alpha phase)	10 20 30	60 62 65	18 20 21	280 310 325	55 65 70
Copper with about 36% zinc	A new structure appears (body centred cubic) in addition to the original face centred cubic structure	3.698 for alpha constituent / 2.935 for new (beta) constituent	Small quantities of a new constituent appear (known as beta phase)	36	70	22	340	60
Copper with 36 to 42% zinc	With increasing zinc more of the alloy consists of the new constituent having body centred cubic lattice	On change to body centred cubic lattice the atoms are 2.935 units apart	The beta constituent increases in quantity while the alpha constituent diminishes	40	85	27	395	45
Copper with 42 to 52% zinc	The lattice is entirely body centred cubic	2.935 with 42% zinc, increasing to 2.941 with 52% zinc	The structure is entirely beta	45	90	28	410	20

* The units of hardness, strength and ductility are explained on pages 98–101, angstrom units on page 296. To convert from angstrom units to nanometres, divide by 10.

3 'vacancies' where atoms are missing;

4 surface and interface defects, which occur at grain boundaries and subgrain or block boundaries.

The micrographs in *Plate 32* show some typical imperfections; they also illustrate how the electron microscope has made it possible to examine metals under resolutions that would have been impossible fifty years ago. *Plate 32c* is an electron microscope picture at a magnification of 50 000. A piece of aluminium has been given a 10 per cent reduction in thickness. The fuzzy dislocations, like tangled wool, are in a cellular arrangement around the white areas, known as subgrains, where there are very few dislocations. Normal shaping operations on such a metal use reductions of 25 to 50 per cent, so we can see how such heavily worked metals contain many dislocations, and why metals become harder on cold working.

Plate 32b which looks like part of a painting by Joan Miró, shows dislocation lines that start and finish at black spots, which are in fact particles of aluminium oxide trapped in a sample of brass. The photograph, at a magnification of 10 000, shows how the oxide particles trap the dislocations. The effect is to harden and strengthen the alloys.

Plate 32c shows an iron alloy containing 3.25 per cent silicon, taken at a magnification of 2200. The three parallel bands are slip bands, which will be discussed again in Chapter 10. The dotted line that crosses the slip bands is a dislocation.

Plate 32d is a micrograph of the iron–silicon alloy referred to above. It has been cold rolled to give a 4 per cent reduction in thickness and annealed until recrystallization commences. The dark areas are those where the effect of dislocations remains; the light areas show that, on annealing, new grains have begun to grow.

In the 1940s, Lawrence Bragg was the first to demonstrate the kinds of disarrangement which a regular pattern of uniform atoms might undergo at a metal grain boundary. He displayed this in a simple and rather beautiful experiment, using bubbles on the surface of water. This is the so-called 'bubble-raft experiment'. *Plate 8* shows the way in which a large group of bubbles arrange themselves. It might be expected that the configuration of the bubbles would consist of absolutely straight lines, but the photograph shows a slight change in direction near the bottom right, and a Y-shaped join at the top left. What can happen with bubbles can happen with atoms; though it must be realized that this is a great simplification of the three-dimensional 'disarrays' which occur in the structure of metals.

If one considers the tree-like formations known as dendrites (illustrated in *Fig. 14* on page 53, it is easy to imagine that a slight bending of each branch while growing from the molten metal leads to a lack of registry when the branches meet and the metal finally freezes. Because the branches are linear, these misfits tend to be in parallel lines between successive branches. Another type of irregularity is a block arrangement known as a subgrain, about 0.001 mm across, where each block of atoms is slightly displaced relative to its adjacent blocks, like a jerry-built brick wall.

A knowledge of the inner structure of metals is of more than theoretical interest, for by helping us to understand how metals and alloys are built up it enables us to exert a more precise and comprehending control over alloy composition and heat treatment than would otherwise be possible. Also, it helps to explain what happens when metals are hardened or stressed. In all branches of science, investigations which at first appeared to be of merely academic interest have often proved to be of great practical benefit, and the study of the inner structure of metals was no exception.

8

SHAPING METALS

How is a sewing needle made? A white-hot slab of 0.8 per cent carbon steel is forced between pairs of rotating, grooved rolls. This process, which reduces the thickness and increases the length of the steel, is followed by a further rolling treatment, using grooved rolls of such a shape that the bar is made into rods of circular section, about 12 mm in diameter. In the next stage the metal, when cold, is drawn through successively smaller holes in hard steel dies which reduce it in diameter, so that eventually it becomes wire of the same diameter as the required needle. The wire is annealed to soften it, precautions being taken that none of the carbon in the steel is lost by the action of the furnace atmosphere. The wire is then cut to a length just over twice that of a needle, and each end is pointed by grinding, in a continuous process, on a rapidly rotating emery wheel. The middle part of the annealed wire is stamped so that it has the form of two needle heads joined together and an eye is then pierced in each head (*Fig. 18*).

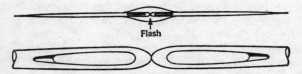

Fig. 18. Needle manufacture: the heads are stamped in pairs

The twin needle is broken into two, and the 'flash' resulting from the stamping operation is ground away. The pieces of wire now begin to look much like needles, although they are still so soft that they can easily be bent double. In the next operation needles are heated to a bright red heat and then quenched. The steel is now extremely hard, and so brittle that a handful of needles can be snapped like uncooked spaghetti. Before they are suitable for use they must be tempered so that, while maintaining much of the hardness, the brittleness is removed. The tempering involves heating to about 200°C.

By now the needles have the correct shape and hardness, but they are scratched and dirty. For many years needle-makers next used what was nicknamed the 'roly-poly process'. Thousands of needles were placed in a canvas sheet, covered with emery powder and soft soap; then the canvas was rolled up and revolved between weighted rollers for many hours. This traditional cleaning process, which was far from ideal for the health of the workers, has now been superseded by one in which about the same number of needles are placed in a vibrating trough, together with oil and emery paste. After about 24 hours the needles are removed, washed clean in soapy water, then dried in sawdust.

Most metal articles have more or less as fascinating a manufacturing history as a needle. Although a multitude of processes are used, they can be grouped into five classes: shaping from molten metal, from hot solid metal or from cold metal (these three are discussed in the present chapter); the joining of metals (described in Chapter 24); and powder metallurgy (Chapter 25).

SHAPING FROM MOLTEN METAL

The remains of an ancient furnace for melting bronze in cup-shaped crucibles, dating from 3000 BC, have been found in Abu Matar, near Beersheba in Israel. Today the same principle is used for holding up to a few hundred kilograms of molten metal. The crucibles are contained in furnaces heated by gas, oil or electricity, and insulated so that as little energy as possible is wasted. When the fuel is gas or oil the outgoing furnace gases preheat the incoming air so that combustion is efficient. For melting large amounts the metal is contained in a shallow refractory bath of comparatively large area. Hot gas or an oil flame plays on the sloping furnace roof and 'reverberates' heat onto the surface of the metal.

During this century several types of electric furnace have been developed. Electric arc furnaces for melting and refining scrap steel are described on page 120. Their high power densities produce melt rates up to four times faster than is achievable in conventional fossil fuel fired furnaces. Electric induction furnaces are clean, efficient and give a lower melting loss than fuel fired furnaces. Their frequency ranges from 50 Hz for very large amounts of metal to 500–1000 Hz for medium-sized melts; in laboratory furnaces for melting small experimental batches, the frequency is as high as 10 000 Hz. An important step in the melting of molybdenum, titanium, zirconium and some 'superalloy' steels has been the perfection of consumable-arc-vacuum and protective-atmosphere furnaces. More recently, beams of electrons have been used to ensure the highest purity of metal during the melting operation.

MAKING A SAND CASTING

Let us assume that a casting is to be made, shaped as in *Fig. 19a*, and that only the simplest equipment is to be used. A solid model or pattern of the letter M is first prepared, and from it an M-shaped cavity is made in moulding sand. A pair of metal boxes, open at the top and bottom, will contain the sand and help it hold together. One box is placed on a board, and the pattern of the letter M is placed face downwards in a central position. The box is filled with sand, which is rammed firmly round the pattern. Another board is placed on top of the moulding box, which is then turned over so that the pattern is uppermost (*Fig. 19b*).

The pattern is now carefully tapped or vibrated, and then extracted from the sand, leaving an M-shaped cavity. It would be possible to make a casting by pouring liquid metal into this open mould, but the top face of such a casting would not be flat, because the metal shrinks as it solidifies. In practice, moulds are made in two parts, and a moulding box, containing sand with a carefully smoothed surface, is placed on top of the first moulding box containing the impression of the letter M. The cavity is now completely enclosed in sand (*Fig. 19c*).

A channel or 'runner' must be cut through the sand in one half of the mould so that liquid metal can be poured down this runner and flow into the cavity (*Fig. 19d*). One or more channels, called 'risers', must also be made; the cast metal enters the runner, fills the mould and rises up the riser, which assists complete filling and helps to ensure soundness in the casting by providing an outlet for air in the mould. When the metal has solidified and the two halves of the mould are separated, the casting remains attached to necks of metal, representing the runner and riser, which are subsequently cut off (*Fig. 19e*).

The description above has been of a simple 'one-off' casting made under primitive conditions. Anyone who has not seen modern foundries at work may be surprised to know that complicated sand moulds can be made which will stand up to the stream of molten metal without being washed away. The choice of the right kind of sand and the strengthening of it with bonding materials make this possible. The sand must be packed or rammed to the correct degree. If it is rammed too tightly, air contained in the cavity, and water vapour from the heated sand, are

Fig. 19. Making a sand casting

trapped in the molten metal, causing unsoundness. On the other hand, insufficient ramming may lead to sand being washed away by the cast metal. Because of the contraction which occurs on solidification, extra molten metal is added after the main bulk has been poured. In casting a marine propeller, to be described later, the 'feeding' continues for twelve hours.

When a casting has to include holes or other recessed features, they are made by separate cores which are fitted into the mould after the pattern has been taken out. Cores are frequently made of sand bonded with oil and baked to give them strength and rigidity, or bonded with chemicals that cause the sand to harden. A complicated mould may have over a hundred cores.

With the advent of automatic moulding machines, more precise control of sand properties became possible. New chemicals have been introduced to strengthen the sands, including 'shell moulding' resins which harden when heated, and new materials in the so-called hot box process, where the moulds are cured in contact with heated metal patterns or boxes. More recently, 'cold curing' resins have been developed. By correct selection of materials and catalysts it is possible to obtain mixtures which either harden in a matter of minutes or, using special dispensing equipment, they may be hardened in seconds.

For the production of large quantities of similarly shaped castings, foundries are mechanized, following the principles of mass production. The patterns are of metal and the moulding boxes are mounted on a conveyor. The sand, having been mechanically mixed and reconditioned, is automatically flung into the moulding boxes which are vibrated on machines so that the sand is consolidated correctly. The moulds are assembled and the metal is poured into them. After the belt has moved forward and the casting solidified, the mould is automatically tilted and the solid casting removed. The mould box continues and the sand falls out to be reconditioned. Under such conditions very high output can be achieved, with rates of production of about three hundred castings per hour.

Components such as cylinder heads for racing cars are subjected to severe working conditions, and the need for the utmost reliability is paramount. Among the worldwide efforts to achieve perfection, the Cosworth casting process, developed in Britain, involves the use of high purity alloys and specially designed moulds. Molten metal is transferred from under the surface of the melt, using an electromagnetic pump to avoid turbulence and oxidation. Zircon sands, bonded and stabilized by chemical hardening processes, are used for the moulds and cores. They are expensive, but are more refractory and consistent than silica sands. A reclamation process was developed to give 100 per cent reutilization of the expensive zircon sand. In the USA, Ford are producing V8 cylinder blocks at a rate of over 90 per hour by the Cosworth process, and report that cost savings have been achieved.

CASTING A LARGE MARINE PROPELLER

Early propellers were made of cast iron or cast steel but, towards the end of the nineteenth century, a copper–zinc alloy (which we now call high tensile brass) containing iron and aluminium was introduced to provide strength and resistance to corrosion. After the Second World War the industry wanted an alloy with still better properties, and now nickel–aluminium bronze containing about 10 per cent aluminium and 5 per cent each of nickel and iron is used for most ships' propellers. Such alloys possess great strength without needing heat treatment, and are very resistant to corrosion and erosion. *Plate 9* shows the propeller for STS *Hellespont Paramount*, an oil tanker of 400 000 tonnes deadweight. The propeller weighs 75 tonnes and is 9.85 metres in diameter; its size is apparent from the small figures of the technicians. This propeller was cast in a proprietary alloy called Nikalium, which is composed of copper with 9 per cent aluminium, 5 per cent iron, $4\frac{1}{2}$ per cent nickel and $1\frac{1}{2}$ per cent manganese.

The material used for propeller moulds is a mixture of silica sand, cement and water. Pits are necessary for moulding large propellers, partly because of safety requirements (should any molten metal escape) and partly to bring the top of the mould to a reasonable height for working. The mould for each blade is made in two halves, the 'bed' and the 'top'. The upper surface of the bed defines the pitch face of the blade, while the lower surface of the top defines the suction surface. The blade centre-line and the approximate shape are marked out. Wooden shuttering is erected to form a box into which the sand/cement mixture is rammed. The pitch face is then formed by what is termed 'strickling': a board with a long arm, pivoted at a central pillar at one end and a roller at the other, rides on an inclined rail, sweeping the sand into the helicoidal surface required for the bed part of the mould.

Next the blade pattern is constructed from templates and the mould top is made, with the incorporation of reinforcing steel. After this has hardened sufficiently the top is lifted, the blade pattern removed and the top replaced, forming the cavity for the blade. This process is repeated for each blade, and a central chamber is moulded which will form the propeller boss. A propeller mould in preparation is shown in *Plate 10*.

After final assembly the mould is clamped securely and held down to the floor with heavy T-bolts. This is necessary because the mould for such a large propeller will contain over 100 tonnes of molten metal, and enormous forces trying to disrupt it will be generated. For some hours before pouring, hot air is blown into the mould to remove moisture. The alloy is melted in large electric induction furnaces. Its composition, gas content and temperature are measured and adjusted.

When the *Paramount* propeller was being made, five furnaces were needed to melt 114 tonnes. The molten alloy, at a temperature of 1250°C, was transferred to ladles, each positioned close to a large runner box whose plug hole led down to the bottom of the propeller mould. The boxes were filled with molten metal and, at a given signal, iron plugs closing the exits from the boxes were released. In the first stage 95 tonnes of molten metal was poured into the mould, which took about 20 minutes. Over the next 12 hours, more molten metal was added to compensate for the shrinkage of the solidifying casting.

A few days after pouring, the mould was dismantled and the casting lifted out. The solidified runners and risers were removed, the propeller's bore was machined, and the blade surfaces were ground to achieve a smooth surface. Finally the propeller was checked for balance, and edge protection was fitted for safe transport and delivery to the vessel. In all, over 35 tonnes of surplus material, including the weight of the runners and risers, was removed to leave the finished weight of 75 tonnes.

Such casting operations make one appreciate the motto on the coat of arms of the University of Birmingham Metallurgical Society: 'the hand that wields the ladle rules the world'.

CASTING RODIN'S *THINKER*

Auguste Rodin's first conception of *The Thinker* was a brooding figure only about 80 cm high, sitting above the bronze doors for a projected Museum of Decorative Arts, inspired by Dante's *The Gates of Hell*. The task occupied him for twenty years, during which he sculpted 186 virile and graceful figures in various degrees of anguish, terror and voluptuousness. *The Thinker* was intended to contemplate the dramatic scenes below. Rodin was never specially interested in the possible culmination of his efforts – the bronze casting of the doors – but several years after his death the gates were cast, based on his models; copies can be seen in Paris, Philadelphia, Tokyo and outside the art museum at Stanford University, California.

Rodin had sculpted *The Thinker*, larger and in a slightly different position from the one above *The Gates of Hell*. The best-known bronze replica of *The Thinker* is over the Paris grave of Rose and Auguste Rodin, but other examples, most of them cast by the sculptor's descendants, are in Cleveland, Pasadena, Baltimore and New York.

The bronze castings are hollow shells formed from clay models of the sculpture. The model would be divided into a front and a back half by inserting a line of 'shims' into the surface of the clay. One half of the model was then coated with liquid plaster about 30 mm thick; when it had hardened the shims were removed and the leading edge of the plaster was treated with varnish so that it would not stick to the next application of liquid plaster, which was then applied to the second half of the clay model. When the plaster had hardened, the two shells were separated and pulled off the clay along the line formerly made by the shims. *Figure 20a* shows the shim line for one of *The Thinker*'s hands.

All of the clay was then removed from the interior surfaces of the two plaster shells. These two halves were then firmly fixed together in their original position, making a cavity which conformed in every feature to the original model. The surface was then given a dressing of liquid soap to make sure that the plaster, which was now poured into the cavity and allowed to harden, would separate easily. The stage had now been reached where there was a plaster replica of the model (*Fig. 20b*).

Next a series of 'piece moulds' were made around the replica. Nowadays foundries use a rubbery composition. In areas of 'undercutting', for example under *The Thinker*'s arm, multiple small pieces had to be made. Each piece was detached from the model and all the pieces fitted together, like a three-dimensional jigsaw puzzle, so that once again a cavity had been created. Molten wax was poured and brushed into the cavity so that the interior was covered with a layer of wax about 10 mm thick. Before this had hardened, metal pins were pushed through the surface of the wax and joined into the material as it hardened. The hollow interior was then filled with a refractory mixture of silica sand and plaster.

The foundry now had an interior core of refractory covered by a skin of wax; this was next covered in a block of refractory mould material as shown in *Fig. 20c*, positioned so that the wax pieces pointed downwards, and then heated. The wax melted away, leaving a cavity mould with a refractory core, ready to receive the molten bronze. As shown in *Fig. 20c*, channels were made in the mould so that the molten bronze would fill it efficiently. The system was arranged so that as the metal rose up the cavity it drove out trapped air, to prevent any faults due to porosity in the casting. The bronze was allowed to cool, the mould was broken

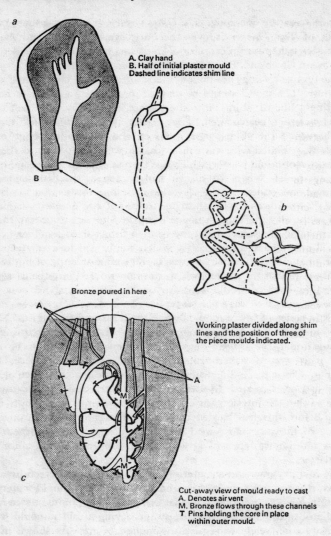

a

A. Clay hand
B. Half of initial plaster mould
Dashed line indicates shim line

B

A

b

Bronze poured in here

A

Working plaster divided along shim
lines and the position of three of
the piece moulds indicated.

A

M

M

c

Cut-away view of mould ready to cast
A. Denotes air vent
M. Bronze flows through these channels
T Pins holding the core in place
    within outer mould.

Fig. 20. Casting *The Thinker*

and the casting trimmed and treated on the surface to provide the required colour and texture. The equally famous statue of Eros (*Plate 20*), cast in aluminium, was produced in much the same way as *The Thinker*.

DIECASTING

Anyone who learns about foundry casting might point out that, though the pattern is used repeatedly, it is a pity that the sand mould has to be made over and over again. This handicap is overcome by diecasting, whereby permanent metal dies are used for making large quantities of castings in non-ferrous alloys. In one process the two halves of the die are made of steel or cast iron; the molten metal is poured into the die cavity, either manually or from an automatically operating ladle. The process is called permanent mould casting, or gravity diecasting, because the metal enters the die under its own weight. In line with other developments, gravity diecasting has been mechanized so that the opening and closing of the die, the operation of cores, which form recessed features, and sometimes the preparation of the die for the next casts are done automatically for castings required in considerable quantity.

Developed at the beginning of the twentieth century, pressure diecasting is today one of the most versatile ways of mass-producing components of great accuracy and good surface appearance. One die half is fixed to the body of the diecasting machine and the other half can be moved backwards and forwards on tie bars, so that it is closed for the casting operation. Molten alloy is injected into the die cavity under high pressure, making a precise reproduction of the form of the component which has been machined into it. Also, and very importantly, sufficient hydraulic pressure is exerted on the closed die for it to remain closed during the fraction of a second it takes for the molten alloy to be injected. The die is provided with cooling channels through which water or oil circulate, to cool the die after casting. After the metal has been injected the die is opened, the design of the component being arranged so that it adheres to the moving half of the die. The casting is pushed away from the face of the moving die half by ejectors, which are steel rods, 3–10 mm in diameter, in suitable positions in the moving die block. Then the diecasting can be removed, either by a robot or by arranging for it to fall into a bath of cooling water, from which a conveyor takes the diecasting to the next operation.

For low melting point alloys, especially those of zinc, the molten metal is held in a container which is embodied in the machine. A plunger, permanently immersed in the molten metal, is forced downwards and this causes a 'shot' of alloy to be injected into the die cavity. Such a machine is called a hot chamber diecasting machine. Zinc alloy diecastings are made automatically on hot chamber machines at speeds depending on the weight and complexity of the components. A casting weighing several kilograms may be produced at the rate of two per minute, and an automobile mirror bracket at five or more per minute, though output is often increased by making a number of parts in the same die. One process engineer supervises a group of machines. Very small parts such as umbrella ferrules and zip-fastener slides are produced at several hundred per minute.

Alloys of aluminium and of copper, which have higher melting points than that of zinc, are diecast in cold chamber machines. The molten alloy is held in a crucible or other melting unit adjacent to the machine. An amount sufficient for one shot is ladled, generally automatically, into the plunger cylinder of the diecasting machine, and a hydraulic ram forces the metal into the die cavity. Magnesium alloys can be diecast in either hot chamber or cold chamber machines.

The development of automatic cold chamber machines for aluminium alloys is now well established, thanks to the use of robot mechanisms and improved methods of lubricating the dies. The molten aluminium alloy is transferred automatically from the crucible to the injection mechanism of the machine; this is done by devices ranging from mechanized ladles to electromagnetic pumps.

For fifty years or more gravity diecasting has been used, not only for medium-sized castings, but also for aluminium alloy components weighing up to 300 kg. So far, pressure diecastings have not reached that magnitude but the diecast cylinder block, weighing about 30 kg (discussed on page 164) indicates the shape of things to come. A cylinder block die weighs as much as 30 tonnes, and the machine is over 15 metres long. Some American, European and Japanese machines have a locking power of over 3500 tonnes, capable of producing complex aluminium components weighing over 30 kg.

A process known as low pressure diecasting has grown immensely in prestige during the last twenty years. It is an economical and efficient means of producing fairly large components in a range of aluminium alloys, some of which are not suitable for pressure diecasting. The die is held above the container of molten aluminium alloy and is connected to the crucible by a ceramic tube. Air pressure applied to the surface of the metal causes a metered amount to rise up the tube and fill the die cavity. Low pressure diecasting has proved especially suitable for the manufacture of automobile wheels.

INVESTMENT, OR LOST WAX, CASTING

In ancient Egypt, before the pyramids were built, craftsmen produced 'lost wax' castings of astonishing beauty and detail in gold, silver and bronze. In modern times the process has been adapted and mechanized, and now plays a very significant role in the aerospace, military equipment and precision engineering industries. The essential feature is the use of an expendable wax pattern. A die containing the shape of the article to be cast made of aluminium alloy is filled with wax injected under pressure to give a replica of the finished casting. Moving parts are incorporated in the die to allow the wax pattern to be removed in one piece. Depending on the size of the finished product, a number of patterns are assembled; for small parts there may be several hundred. In jewellery manufacture high quality rings are cast with about twelve impressions in each mould, while for cheaper rings over a hundred at a time are cast. The buckle of a nurse's uniform will probably be an investment casting made from a mould containing ten impressions. The largest castings, some weighing over 300 kg, are made with only one pattern per mould.

The channel for the molten metal is added, the complete assembly is dipped or 'invested' in a high grade ceramic slurry, and the wet surface is coated with fine refractory particles. This important operation forms the mould coating, which will later come into contact with the cast molten metal. After hardening the coating, the operations are repeated several times on mechanized equipment, using coarser grades of refractory, until a mould is built up of sufficient thickness to withstand the force of the molten metal during the pouring operation.

After drying, the mould is placed in a steam autoclave in which the wax is melted out – hence the term 'lost wax'. The mould is fired at about 1000°C for several hours to strengthen it and to burn off any remaining wax. After the castings have solidified, the mould material is broken away from the cast metal and the runner systems are cut off.

Investment casting has some similarities to diecasting, in that large quantities are especially economical and each product has to be designed carefully to get the best out of the process. Investment castings do not have the disadvantage of the visible joint lines which are inevitable on most castings made from diecasting moulds. The process is widely used for stainless and other alloy steels, for nickel, aluminium and cobalt alloys, and for the various alloys from which permanent magnets are made. Costume and exquisite jewellery are investment cast, coupled with centrifuging during the casting process.

Each year new records are set for the size, complexity and competitiveness of investment castings. In the USA a titanium fan frame hub of an aircraft engine was investment cast, replacing an assembly of 88 small stainless steel castings. The component had a diameter of 1320 mm – a size that would have been impossible a few years ago. Turbine blades in single crystal superalloys (page 48) are now being produced on a large scale. *Plate 5a* shows an investment cast single crystal turbine blade, and *Plate 5b* shows two complex blades for aircraft engines. They were cast in a nickel-based superalloy, with ceramic cores to create the air passages.

SOME OTHER CASTING PROCESSES

In a process called pore-free diecasting, oxygen is forced into the die immediately before casting, thus replacing all the air in it. The oxygen unites with a minute portion of the aluminium or magnesium alloy that has been cast, so that for a fraction of a second the die cavity is virtually a vacuum, and when the metal is cast it will contain no porosity. The oxide content of the metal is so small that its properties are not affected. In Japan in particular, many components of motor cycles are made in this way, and several hundred thousand automobile wheels are produced by pore-free diecasting each year.

In vacuum diecasting, which has been the subject of much research, the aim is to create a vacuum in the die before the molten alloy is injected, thus eliminating the possibility of porosity. Ideally a 100 per cent vacuum would solve all the problems, but in practice a low vacuum of about one-fifth of atmospheric pressure can be achieved, and this significantly reduces microporosity.

Squeeze casting is another development that has aroused interest in recent years. In one method the dies are held in a vertically operating press and the molten metal is poured directly into the impression in the bottom die. The top die half is then closed on to the bottom die half, subjecting the metal to great pressure and forcing it to take the shape of the cavity formed between the two die halves. This and similar methods are suitable for the usual aluminium casting alloys, and also for alloys normally shaped in forging presses. Squeeze casting is used to produce automobile wheels and heavy section components not suitable for pressure diecasting.

CASTING SEMI-SOLID ALLOYS

When molten alloys solidify, most of them pass through a semi-solid stage (page 39). It was found that alloys in that condition could be stirred vigorously by electromagnetic means and cast into a mould just as if they were completely liquid. On account of the temperature being lower than that required in conventional casting methods, castings of great solidity can be made. Furthermore, non-metallic fibres can be incorporated in the cast metal to give increased strength. Several large companies, mostly in the automotive industry, are now cooperating to select suitable components to produce prototypes and evaluate possibilities.

CENTRIFUGAL CASTING

For casting drainpipes and similar shapes, a metallic cylindrical mould, without any cores, is spun at high speed. Liquid metal is poured into it, so that centrifugal force flings the metal to the face of the mould, thus producing a cast hollow cylinder of uniform wall thickness. This process yields a product having a dense, uniform outer surface; consequently a cylinder liner cast by this method is considered superior to similar ones cast in sand moulds. Cast-iron piston rings are cut from such cylindrical shapes made by centrifugal casting, and the process is now being used to make complicated components.

CONTINUOUS CASTING

Until the mid-1930s molten metal would be cast into ingot moulds ready for remelting, or slabs would be delivered to rolling mills where the metal would again have to be reheated. Then, from small beginnings, the process of continuous casting started on its development into a major industry, culminating in the continuous casting of steel. First, the basic principles of the process for non-ferrous metals are described.

A water-cooled copper or aluminium mould about 150–250 mm deep, having the shape of the desired cross-section of the slab to be cast, is sealed at the bottom by a retractable base (*Fig. 21*). Molten metal is poured into the cavity continuously while the baseplate is slowly lowered; the metal solidifies, first as a shell adjacent to the mould and then, as solidification progresses, into the centre of the cavity. The process is hastened by additional cooling by water jets and by withdrawing the billet into a tank of water below the mould. Almost all aluminium alloys for rolling are now semi-continuously cast; a length of about 3–6 metres is poured, casting is stopped and the solid metal is withdrawn. Casting is recommenced by returning the base to its original starting position.

This development was relatively easy for metals melting at up to

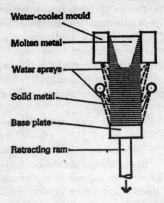

Fig. 21. The principle of continuous casting for aluminium

700°C, such as aluminium, but in the past thirty years the immense problems associated with the continuous casting of steel have been splendidly overcome. Today the continuous casting of steel is a large and flourishing industry, with an output of over 400 million tonnes a year. It is expected that at the end of the century about 90 per cent of all steel made will be continuously cast. (The methods are discussed on page 123 and illustrated in *Fig. 37*, with a picture of a modern plant in *Plate 17*.)

THE SHAPING OF SOLID METAL

If a bar of cold steel is hammered, a great amount of energy is needed to change its shape permanently. The same metal, when heated to bright redness, is softer and more pliable. The village blacksmith makes use of this and, with his hammer and forge, shapes hot steel into horseshoes and an abundance of tools and fittings.

With the coming of the machine age, mechanical hammers were designed and constructed which could forge larger shapes than the black-smith was capable of tackling. An example of one of these old machines, the tilt hammer, can be seen at the Abbeydale industrial hamlet in Sheffield. An iron hammer-head weighing about 50 kg is fixed at the end of a long wooden arm, pivoted at the centre. A cam rotates beneath the arm, so that the head rises and then drops by its own weight onto the piece of hot metal held beneath it.

A development of the same principle is the drop hammer, in which a heavy steel die block between two vertical guides is mechanically lifted about one and a half metres above the anvil and allowed to fall under its own weight onto the metal to be forged. Two halves of a die are made: one block is keyed to the anvil at the base of the machine, and the other is fixed to the drop hammer. A hot bar of metal is firmly held on the anvil by tongs, and the hammer falls to forge the metal between the two halves of the die.

For a fairly complicated part, the die may contain three or more pairs of impressions of progressively increasing detail, the purpose of which is to change the shape of the metal in gradual stages. Thus multiple forging operations are performed with the same die, and for each of these the workman places the piece of metal in the various positions during successive blows of the hammer. The first die causes the metal to assume a form roughly approaching that of the finished article, the next brings the metal practically to the desired shape, and the final die impression makes the forging accurate in its dimensions.

In a modification of drop forging, mechanical or hydraulic force is used to push the hammer downwards, thus increasing the power of the blow and the speed of the operation. Automation features in the mass production of forgings; water-cooled punches and dies perform up to four operations simultaneously, producing flash-free components. *Plate 11* shows a 6000 tonne mechanical press for producing diesel engine crankshaft forgings up to 60 kg.

For very large forgings such as marine engine crankshafts and propeller shafts, and aerospace components, the forging is done by hydraulic presses which generate pressures of up to 60 000 tonnes. These presses, though often several times larger than mechanical presses, are much quieter in operation.

EXTRUSION

The process of extrusion works on much the same principle as squeezing toothpaste from a tube. A prodigious pressure would be necessary to extrude most of the common metals while cold, so plant is generally designed to extrude hot metals. For example, copper is extruded at temperatures between 800 and 900°C, cherry-red to orange; copper alloys between 600 and 800°C; and aluminium alloys between 400 and 500°C, just visibly red. At the other extreme, some steels are extruded at white heat, 1200 to 1400°C.

A modern extrusion press used in the production of copper tubular products will take a block of cast copper in the form of a cylinder about 300 mm in diameter and 660 mm long, weighing about 450 kg. This is heated to 850°C and then placed in line with the container of the press. One ram pushes the metal into the press, and a second ram pierces the centre of the billet with a mandrel. The main ram exerts a high pressure, forcing the copper between the mandrel and a die, to produce a long hollow shell. Thus within 15 seconds the metal block is transformed into a tube, typically 75 mm bore by 3 mm wall thickness and over 40 metres long.

The principle of this type of extrusion press is illustrated in *Fig. 22a*; the process is known as 'direct' extrusion. Another method, known as the 'indirect' process, is illustrated in *Fig. 22b*. This has some technical advantages, such as less power required and greater uniformity of structure of the metal, but it requires a more intricate design of machine. The main difference between the two methods is that in the direct process the metal is forced through the die, while in the indirect process the die is forced through the metal.

Die orifices are made in a variety of shapes to produce different sections such as curtain rails, and windscreen and window sections for automobiles. To make gear wheels for small machinery, the continuous gear-shaped length is first extruded and is later sectioned into gears. The process is so economical that for copper, aluminium, magnesium and lead alloys extrusion is a normal production procedure for such products as rods, bars, tubes and strips. The extruded rod or section is sometimes finally drawn through another die in the cold state, in much the same way as for wire drawing (*Fig. 27* on page 88). This is in order to improve the dimensional accuracy, and often to increase the strength of the alloy.

Extrusion can also provide a starting point for the production of tubes and pipes, since by modifying the press it is possible to arrange for the metal to be squeezed between a solid steel mandrel and a die, as shown in *Fig. 22c*. As an example of the versatility of the process, lead tubing can be extruded and shrunk directly on to finished cotton-covered insulating wire, thus forming lead-covered cable for underground telephone lines.

ROLLING

Metals may be rolled hot or cold. The advantage of hot rolling is that most metals can be reduced in thickness much more easily than when cold-rolled. On the other hand, the surface finish and accuracy of cold-

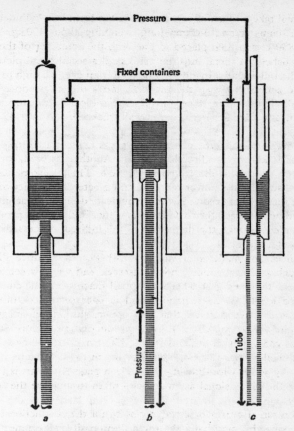

Fig. 22. Three types of extrusion

rolled metal is better. Ingots of steel used to be transported to the rolling mill, where they were reheated to white heat. Then the ingot would be moved to a 'cogging mill' with a number of grooves in the rolls. The steel would be passed backwards and forwards through the rolls, travelling through different pairs of grooves at each pass, so that it was gradually reduced in thickness. Then the rolled steel would be reheated and further reduced in thickness by subsequent rolling. The finished product would be long lengths of rod or bar,

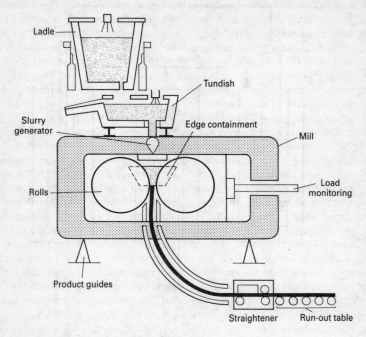

Fig. 23. Direct casting of steel strip

the steel having been sheared during several stages of the rolling operation.

During the past decade, thanks to the progress made in continuous casting, the production of rolled steel products has been revolutionized. Most major steelmaking companies have been developing near-net shape casting of strip (1–5 mm), which eliminates the strip mill, and of thin slab (30–60 mm), which eliminates the roughing or cogging mills. Direct strip production based on continuous casting, illustrated in *Fig. 23*, is achieving a substantial reduction in processing costs.

Nowadays the operations of continuous casting and final rolling are combined with precise control of the metal's structure. The modern railway line calls for such control; trains are running faster, axle loadings are increasing and tracks are more intensively used. British Steel, the world's largest exporter of rails, has developed a process in which the hot rail passes from the finishing stand at about 1000°C. The temperature profile along its length is measured, and the data are used to program individual water sprays along a cooling section, where rollers maintain the straightness of the rail as it cools. In this way the rail leaves the unit at a uniform temperature, the structure – and therefore the hardness – having been precisely controlled. Rails with a typical composition of 0.80 per cent carbon and 0.85 per cent manganese (balance iron) are formed with a hardness of 290 Brinell; service lives four times those of earlier rails have been obtained.

The power required for cold rolling is greater than for hot rolling, and the requirements for accuracy and surface finish are more precise. Most cold-rolling mills producing steel strip and sheet are known as tandem mills, and consist of 2, 3, 4 or 5 stands about 4 metres apart. In order to provide the required accuracy in the finished product, the working rolls are backed up top and bottom by other rolls at least three times the diameter of the work rolls; their function is to support the work rolls, preventing them from distorting. As the metal is progressively reduced in thickness, the speed of the rolls must be accurately controlled in order to avoid buckling or breaking of the steel. This is done by small 'tension rolls' over which the strip passes between stands. If there is any tendency to buckling, the tension roll records the drop in pressure and the setting of the work rolls is rapidly corrected. The loading or squeezing power of the rolls, the adjustments for controlling the thickness of the strip, the lubrication, and the mechanism which directs the metal from one stand or mill train to the next are all controlled automatically. The rolls and their housings and bearings are designed so that in case of breakdown rapid replacement is possible. The work rolls, weighing a few tonnes, are changed regularly, which takes about 15 minutes. Occasionally, when the backing rolls and housings need changing, this involves moving over 50 tonnes of metal.

One cold-rolling mill in the USA accepts strip up to 5 mm thick and from 100 to 180 mm wide, weighing up to 40 tonnes, and cold-rolls it down to 0.15 mm thick. The mill has five sets of rolls in tandem, and as the strip passes from one to another it gathers speed till the last set of rolls is delivering sheet at 1500 metres per minute. All the sequences of operation, the tracking of the coil of metal, and the regulation of the rolls are computer-controlled, while closed-circuit TV is used to assist supervision (*Plate 12*).

As cold steel is harder than hot steel, cold-rolling mills need harder rolls and, as mentioned above, the work rolls have to be supported to prevent distortion. In the Sendzimir cold mill, illustrated in *Fig. 24*, the tungsten

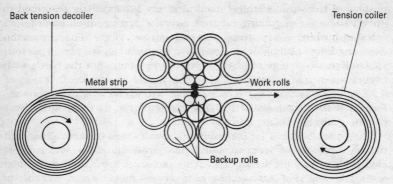

Fig. 24. Rolling strip in a Sendzimir mill

carbide work rolls are backed up by a cluster of heavier rolls. This type of mill is in wide use, especially for rolling stainless steel. One advantage is that roll changes take only a few seconds since the small, hard work rolls can be withdrawn quickly from this position and replaced by newly reground rolls. As in all cold-rolling mills, the Sendzimir rolls are flooded with a coolant while in operation to prevent the tremendous surface pressure damaging of the rolls and the strip.

TUBE-MAKING

Over a hundred years ago, steel tubes began to be used in making bicycles. The Singer Xtraordinary and the Kangaroo, both penny-farthings built in 1878, contained steel tubes to join the handlebars to the small rear wheel. The great modern tube-making industry derived from these beginnings and, though the products of the tube-makers' skills are evident in a thousand industries, a close connection has been retained with cycle-making. Nowadays the welded steel tubes of bicycles are shaped and manipulated, to be thick where the stresses are greatest and thin where the stresses are least, so that they possess the correct combination of strength, flexibility and lightness. After cycle manufacture demonstrated the value of tube construction, their use spread to motor cycles; now tubes feature in products ranging from hypodermic needles (famous as the smallest tubes ever made) to land and marine boilers, hydraulic piping, axles of cars, trucks and railway wagons, and to many of the things we use at home or in sport – prams, furniture, javelins and millions of golf-club shafts.

One of the world's largest manufacturers has modestly described the two basic ways of making tubes as either wrapping an accurate hole in steel or pushing a very strong hole through a steel bar. The first method was used for making lead pipes two thousand years ago, and today a modern derivation of that process makes untold metres of electric resistance-welded tubes.

The process begins with coils of steel strip fed automatically through a series of forming rolls which gradually curl the flat strip into a tube shape. The edges of the tube are heated by electric induction and pressed together so that they join by welding. Then the tube travels through a series of rolls which size and straighten it. Although many such tubes are supplied in straight lengths, a great number are manipulated – bent, tapered, flanged, swaged, screwed, tapped and slotted – into the shapes required by manufacturers.

The second type of process, where the 'strong hole is pushed through a steel bar', has many variations, but the products are generally described as seamless tubes. As long ago as 1890 the Ehrhardt process proved ideal for producing tubular items which required a good solid strong end, such as high pressure gas cylinders. This process was developed and automated until today millions of gas cylinders have been made under quality control conditions of the utmost severity, aided by ultrasonic testing and continuous appraisal throughout the process, starting with the chromium–molybdenum alloy steel bar and finishing with the gas cylinder. A permanent record is kept of every cylinder made in the last eighty years. *Figure 25* illustrates the stages in the manufacture of a typical high pressure gas cylinder.

The Mannesmann process for making seamless steel tubes is illustrated in *Fig. 26*. A solid rod of hot steel is spun between two mutually inclined rolls which rotate in the same direction, so that the rod is pulled forward between them and passes over a mandrel. In this way a thick-walled tube is produced, the dimensions of which can be varied by setting the rolls and the size of the nose-piece of the mandrel. After that, the tube-shape can be fabricated into a thin-walled tube by further processing.

In a derivation of the Mannesmann process, a piercer with three rolls provides greater accuracy and concentricity than was possible with the Mannesmann mill. Round steel bars are cut to length by automatic oxypropane burners. Each piece is check-weighed and transferred by a conveyor to a large furnace with a rotating hearth which passes through several temperature zones in the furnace, finally heating the steel to 1250°C. The hot billet is passed to a hydraulic press which makes a cone-shaped indentation in the centre of the billet's end, making it ready for the piercing operation which follows.

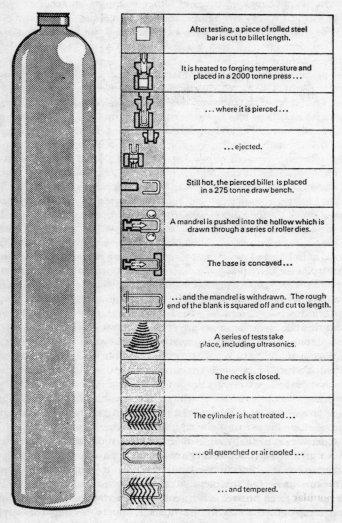

Fig. 25. The making of a high pressure gas cylinder
(*Courtesy Chesterfield Cylinders Ltd*)

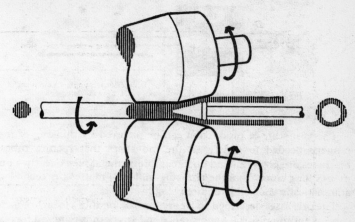

Fig. 26. Mannesmann tube-making process

Next, each billet passes through a three-roll piercer where three barrel-shaped rolls grip the billet and force it over a conical plug in a similar but more controlled way than in the Mannesmann two-roll process. This produces a thick-walled short tube, usually called a 'bloom' in the tube-making industry. A mandrel bar is fed into the bloom, and both are passed through a series of eight stands, each consisting of a pair of grooved rolls. Each succeeding pair of rolls reduces the diameter and thickness of the bloom, while the bore is supported by the mandrel. A typical tube formed at this stage of the process would be 20–25 metres long. The next stage converts this into a finished tube three or four times as long on a 'stretch reducing mill', consisting of a series of 24 stands, each with a cluster of three grooved rolls which progressively reduce the diameter while increasing the length of the tubes. The speed of each roll cluster is set individually to enable the stretching action to take place between each stage. The tubes are examined for surface defects by visual and magnetic crack-detection techniques.

A process known as 'pilgering' or 'tube reducing' is becoming increasingly popular. The process is normally used in seamless tube production where an extrusion press produces a shell whose wall thickness may be up to a third of the outside diameter. Two grooved rolls, which are given a reciprocating movement, embrace the tube from above and below so that the diameter and wall thickness of the tube are rapidly reduced.

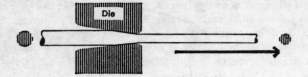

Fig. 27. Wire drawing

The vast scale of production can be pictured by thinking of the large tonnage needed to satisfy all the industries that require tubes. The machines are powerful and operate rapidly; metal tubes nearly a hundred metres long come from the reducing mill at 4½ metres per second and are automatically cut to the required lengths.

The processes described so far are for making tubes from hot metal. If a tube of optimum surface appearance and strength is required for, say, aircraft structures or hypodermic needles, a further process of cold drawing is applied. The tube, pointed at one end, is inserted into a die and gripped by a movable carriage which pulls the tube through the die and over a plug which has been screwed into position on a mandrel. This has the effect of reducing the diameter and the wall thickness of the tube and increasing its length. The cold-drawing process is used not only for ordinary steels, but also, for example, for high strength alloy steel tubes for racing cars and bicycles, and in specialized equipment such as mountain rescue cradles.

THE DRAWING OF WIRES AND RODS

Wires are produced as shown in *Fig. 27*, and are always made at a comparatively low working temperature. The raw material for making steel wire is hot-rolled rod about 6 mm in diameter. After annealing, cooling and removal of scale, the rod is pointed at one end and then inserted through a hole in the die slightly smaller than the size of the rod. The pointed end is gripped from the other side and the rod is pulled through, thus reducing its diameter.

In modern continuous methods, particularly as applied to copper, the wire is drawn through a first die, turned round a roller, then passed through a second die, and the same procedure is repeated several times, using successively smaller holes and rollers of increasing peripheral speed. By the time the wire enters the last die it is much reduced in diameter and is travelling at high speed. In order to save continual rethreading, lengths of rod are welded together before drawing.

The annual world consumption of steel rod for wire drawing is over 30 million tonnes. Substantial quantities of wire in other metals are also produced. Some is for relatively undemanding applications such as the copper wire used in the home, but a large tonnage is made for purposes that require great strength. British Steel produces about 60 000 tonnes of wire rope per annum, 40 000 tonnes of prestressed concrete wire and 20 000 tonnes of wire for tyre cord and bead. Wire for the reinforcement of tyres is made of 0.8 per cent carbon steel and is usually around 0.2 mm thick. At the other end of the 'thickness spectrum', wire for suspension bridges is 5 mm in diameter. Bridge designers are now calling for a strength of up to 1770 N/mm² (about 120 tonnes per square inch); this has led to the use of microalloyed steels, which will be discussed on page 145. The major change in the processing of such wires has been the adoption of continuous casting followed by controlled cooling in the rod mills.

SHEET METAL WORKING

The mass production of sheet metal can be traced to the middle of the nineteenth century, when sheets of tinplate were stamped to make food containers. Further rapid growth of this important industry came when automobiles began to be mass produced and domestic equipment, from toasters to washing machines, was required universally.

Low carbon steel is the principal raw material. The process begins with a blanking operation in which huge coils of metal strip, often weighing thirty tonnes, are automatically unrolled and cut to the required size by shearing dies. Then the blanks are stamped in a pair of metal dies which have been shaped three-dimensionally to the required form and which are mounted on great hydraulically or mechanically operated stamping presses. The next step is flanging, in which the metal is bent to provide the areas which will be used to fasten the stamping to other components. This is often followed by a final stamping operation to sharpen the contours of corners.

The modern stamping line is highly automated, with mechanical fingers for feeding and transferring the metal, lubricant sprayers synchronized with the operation of the press, conveyor systems, and an automated inspection system which functions without holding up the sequence of manufacture. Entire automobile panels are now being made in one operation by prewelding sheets with high speed lasers. By welding together steel sheets of different thickness and composition, automotive engineers can design the bodywork so that maximum strength is provided in those areas that require it most.

OTHER METHODS OF SHAPING METALS

Metal buttons for uniforms are made from metal strip in a number of stages: circular blanks are punched from a strip of annealed brass or nickel–silver, and each blank is placed in turn on the anvil of a stamping machine. One half of the die is mounted on a heavy steel block which is raised to a given height and dropped onto the metal blank, thus stamping the pattern on the front of the button. The back of the button is stamped separately into the form of a shallow cup, and the shank which forms the eye is inserted. In shaping both the back and front a lip is made at the edges and the button is completed by forcing the lip of the back of the button underneath that of the front.

Collapsible toothpaste tubes and patent medicine containers are made by the process of 'impact extrusion'. In this method, akin to the extrusion process described on page 79, a small blank of cold metal is punched in a die. The metal is squeezed between the die wall and the punch, producing a hollow thin-walled container. Some aluminium teapots and hot water bottles are made from sheet metal by 'spinning'; nails are made from wire by a continuous process for forming the heads and points; cartridges are made by a 'deep drawing' operation.

MACHINING OF METALS

All metals can be cut, but there are great differences in their resistance to cutting; for example, it is much easier to cut an aluminium milk-bottle top than a thin razor-blade with a pair of scissors. It does not follow that a soft metal can be machined easily at high speed by turning it in a lathe. The soft metal being cut may build up on the cutting edge of the tool and interfere with further cutting. Faults in the manufacturing process may form inclusions which lead to rapid wear of the cutting tool tips. For example, if an aluminium alloy has been overheated when molten, aluminium oxide is formed, which causes hard spots in the solidified casting. These play havoc with cutting tools, especially in rapid automated machining processes. If an aluminium–silicon–copper alloy containing more than limited amounts of iron and manganese is 'stewed' in a crucible when molten, a heavy sludge is formed which sinks to the bottom of the crucible. When the alloy is cast, some of this sludge may mingle with the metal, causing hard spots which will cause trouble in machining and rapid deterioration of the cutting-tool tips.

To ensure high output and rapid machining, some alloys are made with added elements. Small, rounded, low melting point constituents evenly distributed through the metal assist rapid machining. Well-known examples are steels containing lead, or manganese sulphide particles, and copper and brasses containing lead, selenium or tellurium particles.

The shaping of metals on lathes, drills, millers and many other types of machine is an industry which ranges from the home workshop to the giant automated plants of the motor manufacturers. Massive forging presses shape and manipulate steel ingots weighing 30 tonnes or more, and the forgings are then machined to make drum winders for coal mining, ships' propellers or equipment for steelmaking plant. The machines on which such parts are brought to their final shape are correspondingly immense; for example, lathes 30 metres long have to be accommodated. At the other end of the scale, tiny holes for wrist-watches are drilled with laser beams to obtain the greatest possible accuracy. Ever since metals were first employed, engineers have been devising new ways of shaping them, and there are still no signs that our technical resourcefulness in this respect is being exhausted.

9

COINAGE

Although precious metals were used for trading more than four thousand years ago, it was not until the seventh century BC that a king in the province of Lydia, now part of Turkey, issued coins produced from pieces of gold and silver, stamped with his royal symbol. In the Middle Ages gold and silver coins were made by melting the metals in charcoal-fired furnaces and pouring them over flat surfaces to solidify. The sheet so produced was beaten to the required thickness, and blanks were clipped to the correct size. These blanks were struck between two dies, giving the coin the head of the monarch on one face and an emblem on the other.

In King Alfred's reign there were several mints, but the one at the Tower of London became pre-eminent in the thirteenth century. At that time coins were produced by casting rods of metal and slicing them into coin blanks. When rolling mills were developed in the sixteenth century, they were used to make strip from which coins were produced to a thickness more accurate than was achievable before. One seventeenth-century rolling mill was powered by four horses; hand-operated presses produced blanks from the rolled strip.

In 1810 the British Royal Mint commenced operation in its new premises at Tower Hill, equipped with steam-powered rolling mills designed by Matthew Boulton. He also invented coin presses capable of striking coin blanks at the rate of one per second; they were equipped with devices for automatically locating the blanks between the dies and ejecting the coins after striking. By the end of the nineteenth century, annual production had risen to 100 million coins. Electric power later replaced steam, but the production route of casting the metal, rolling it into sheet, blanking, annealing and coining remained unchanged.

By the late 1960s the production capacity at Tower Hill had become overstretched, and the equipment was outdated. It was decided to build a new mint at Llantrisant in South Wales. The new plant was designed to process copper alloys and to use the latest methods of continuous casting production. Two electric furnaces are used, the larger with a capacity of 3½ tonnes in which the alloys are melted for transfer to the smaller holding furnace. Graphite moulds are positioned at the throat of the furnace and cooled externally by water jackets. As liquid metal enters the moulds, it solidifies and a continuous solid strip is withdrawn automatically.

The Llantrisant Mint has four twin-strand casting lines, producing about 20 000 tonnes of strip a year. The advantage of this continuous casting process is that hot rolling is not required, and it is possible to produce the strip to the required thickness without any intermediate annealing, as had been necessary when the strip was rolled. The presses now produce up to 10 000 blanks a minute from the cast strip, and the coining presses strike up to 700 pieces per minute. Robots, such as the one illustrated in *Plate 13*, are used to stack the coins. The total annual output at Llantrisant is about 2000 million coins, destined for Britain and many other countries.

COINAGE ALLOYS

Economic vicissitudes have caused continual changes in the alloys used for coinage. Until the seventeenth century most coins were of gold or silver, with a few copper coins for low values. In the 1770s, copper pennies and twopences were made in Boulton and Watt's foundry in Birmingham; these coins were so large that they were nicknamed 'cart-wheels'. The use of bronze instead of copper was one of the consequences of the French Revolution. The revolutionary atheists destroyed churches and then tried to find a market for the bronze bells. It was found that when an equal weight of copper was added to the bronze, a good coinage alloy was formed. During the years after 1793 more and more copper was added to the bell bronze until, half a century afterwards, the coinage alloy consisted of 95 per cent copper, 4 per cent tin and 1 per cent zinc; this composition was adopted not only in France but in many other countries, including Britain (in 1860).

During the twentieth century tin became more and more expensive, and when in 1942 Japan occupied the major tin-producing countries, the percentage of tin was reduced to only 0.5 per cent. The changes in the composition of British bronze coins from 1860 to 1992 are given in *Table 6*. Silver coins had been minted by Offa, King of Mercia, in the eighth century, and the metal remained in circulation until 1947. However, for 25 years before that the silver content was being reduced, as shown in *Table 7*.

Table 6. Composition of British bronze coinage

Period	Copper (per cent)	Tin (per cent)	Zinc (per cent)
1860–1923	95	4	1
1923–1942	$95\frac{1}{2}$	3	$1\frac{1}{2}$
1942–1945	97	$\frac{1}{2}$	$2\frac{1}{2}$
1945–1959	$95\frac{1}{2}$	3	$1\frac{1}{2}$
1959–1992 (including decimal coinage)	97	$\frac{1}{2}$	$2\frac{1}{2}$

Table 7. Composition of British silver coinage

Period	Silver (per cent)	Copper (per cent)	Nickel (per cent)	Zinc (per cent)
to 1921	92.5	7.5	0.0	0.0
1922–1926	50.0	50.0 (various silver alloys were used in this period)		
1926–1947	50.0	40.0	5.0	5.0

When it became necessary to replace silver, preferably with an alloy of similar whiteness, several alternatives were considered. Nickel was an obvious choice and had been used by France in 1914 and by Canada in 1922. However, the cupro-nickels became the most popular choice for 'white' coinage. The first coins in this alloy had been made by the Royal Mint for Jamaica in 1869. The composition then was 80 per cent nickel and 20 per cent copper, but now a 75/25 composition is preferred. The present 'silver' coins in Britain are of that composition, as are the US 5 cent 'nickel' and the Indian 1 rupee piece. In addition to such familiar coins, the Botswana 1 pula and 50 thebe, the Brunei 5 to 50 sen coins and the Zambian ngwee are of cupro-nickel.

Stainless steel with 17 per cent chromium has been used as a coinage alloy, but it is difficult to strike in the coin presses. Italy was the first country, in 1939, to employ stainless steel; other countries including Brazil have used the same material. During wars and other times of shortage mild steel has been adopted for coinage; Poland, Germany and Norway used steel during the First World War. To reduce corrosion problems a variety of coatings have been applied, for example zinc-coated steel was used in the USA in 1942, Bulgaria used nickel-coated steel in 1943 and Canada used chromium plated steel in 1944. Cupro-nickel-coated steel was circulated in Argentina in 1945. Steel blanks electroplated with nickel were produced in Canada for El Salvador and Costa Rica in 1977.

Recently, the British 1 and 2 pence coins have been produced in copper-plated steel instead of the bronze alloy. Indeed, there is now such a rapid movement towards electroplated steel that copper alloys are being displaced from their pre-eminence as coinage materials. The Royal Mint at Llantrisant is still producing copper alloy coins, using continuous casting as described on page 93. But they now also buy in continuous cast steel, which is brought to the right dimensions and then plated.

The British 1 pound coin, introduced in 1983, is made of copper with 25 per cent zinc and $5\frac{1}{2}$ per cent nickel. This composition was chosen to enable the coin to have a distinctive electrical conductivity so that it can be used in vending machines. This has now become important, since untold millions of coins are required for electronic machines, from one-arm bandits to parking meters.

Brass, usually with a small addition of tin, has been used occasionally. Aluminium bronze containing 6 per cent aluminium and 2 per cent nickel was represented by the now discontinued British threepenny coin, but France and some other countries have also used this alloy for coinage. Recently composite coins have been produced, known as 'ring and disc' coins. Such bicoloured coinage was an Italian innovation: their attractive 500 lire coins were issued in 1982, consisting of an aluminium bronze disc set in a ring of stainless steel. Other countries circulating bimetallic coins include France, Portugal, Switzerland and Thailand.

The advent of the cashless society has been heralded. Certainly many transactions are now made by cheque, credit card or electronic means, but for minor purchases coins will continue to be required for many generations to come. It is reckoned that countries where gambling casinos are popular require well over a hundred coins per head of population.

10

TESTING METALS

Metals are tough; they are employed where high stresses and strains have to be endured. A steel rod 30 mm in diameter – just over an inch – can support a load of over 25 tonnes without fracture. *Figure 28* attempts to convey some idea of the load which could be borne by such a slender piece of steel.

Different kinds of stress may be experienced in service; for example, the hauling rope, made of stranded steel wire, supporting the cage in a mineshaft is subjected to pulling, or 'tensile' stress, while the vertical columns supporting a bridge suffer mainly compressive stress. Almost all moving parts in machinery undergo rapid changes or combinations of stress; for example, the connecting rod of an automobile is subjected to alternations of tensile and compressive stresses, while the axle of a railway carriage suffers a combination of bending and twisting.

One task of the metallurgist is to specify suitable alloys which will endure the stresses encountered in service, and metals must be tested so that their mechanical properties, especially their strength, can be assessed and compared. When a metal is selected for service, mechanical tests must be carried out as an inspection routine in order to make certain that, throughout the production of batches of that metal, the quality is maintained. The principal method is to test to destruction representative samples of the metal, thus subjecting them to much more severe conditions than they will normally endure. The designer estimates the likely stresses in a part and, from a knowledge of the mechanical properties of the metal, is able to calculate the shape and size required. There is added a safety factor which varies according to the type of stress to be endured and the relative importance of safety so that, for example, a component of a nuclear power station would be given a much higher safety factor than a part for a domestic washing machine.

Fig. 28. The strength of steel
(*Courtesy Joanna Coudrill*)

THE NEWTON

Tensile strength equals force per unit area; engineers in Britain have usually stated the strength of metals in units of tons (force) per square inch, abbreviated in practice to 'tons per in². In America tensile strengths are given as pounds (force) per square inch. These particular conventions, though still widely used, can lead to confusion in thought between the the pound as a mass unit and the pound-force (which is the *force* exerted on a pound *mass* by gravity). A similar situation exists with the metric system of units: the mass unit is the kilogram (kg) and the force unit is the kilogram-force (kgf), and here again the distinction is often blurred in practice.

With the development of metrication the opportunity was taken to adopt the 'Système International d'Unités' (for which the abbreviation is SI in all languages). This is gradually gaining acceptance and is being legally adopted in many parts of the world. In SI the force unit has a distinctive name, the newton, commemorating the great scientist Sir Isaac Newton and his work on the force of gravity. The newton (symbol N) is the force required to produce unit acceleration on a mass of one kilogram.

From the well-known equation Force = Mass × Acceleration, it can be seen that the force required to produce gravitational acceleration (9.807 metres per second per second) on a kilogram mass is

$$\text{Force} = \text{Mass} \,(1\,\text{kg}) \times \text{Acceleration} \,(9.807)$$

The force of 9.807 newtons is equal to the kilogram-force in the metric system. A tensile strength previously expressed as 1 ton (force) per square inch becomes, in SI, 15.44 newtons per square millimetre. The derivation of this conversion factor can be seen from the following:

$$\frac{1\,\text{ton (force)}}{(1\,\text{inch})^2} = \frac{1016\,\text{kilogram (force)}}{(25.4\,\text{mm})^2} = \frac{1016 \times 9.807}{645.16} = 15.44$$

Therefore one ton-force per in² equals 15.44 newtons per mm². When kilograms per mm² are converted to newtons, the factor 9.807 is used.

Table 9 on page 112 gives the mechanical strengths of various metals in tons per in² and in newtons per mm². A similar comparison was included in *Table 5* on page 61.

THE TENSILE TEST

A test-piece is machined to a shape shown in *Fig. 29*. The dimensions of the piece depend on the circumstances of the test; it may be only about 5 cm long, or it may be 50 cm long. The ends will be gripped in the testing machine. Two marks are scribed on the narrower section; they define the gauge length, and will be used for measuring the amount of stretch which the metal will undergo.

The gauge length depends on the dimensions of the test-piece. In Britain this dimension is calculated from the formula 'gauge length = 5.65 times the square root of the cross-sectional area of the central part of the test-piece'. Thus, if the cross-sectional area is 150 mm², the distance between the two marks would be 5.65 × 12.24, or about 76 mm.

The test-piece is placed in a tensile testing machine, which is capable

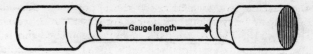

Fig. 29. Test-piece for the tensile test

of applying a steadily increasing and measured pulling force, thus recording the tensile strength of the metal. *Plate 14* illustrates a universal testing machine as used in test facilities in many parts of the world. It measures the tensile or compressive strength of metals, and is programmed so that results can be stored for comparison.

The stress is applied gradually but, though the recording dial soon shows that a 3000 kg force is being exerted, the steel appears unchanged. If, however, the distance between the two marks – the gauge length – is measured with an extensometer while the steel is still under tension, it is found to have increased slightly, by less than a tenth of one per cent. If now the force of 3000 kg is removed, the steel returns to its original length. In other words, the metal has so far behaved elastically. The relationship between the applied stress and the amount of strain (in this case the extension of length) is known as the modulus of elasticity.

When the tensile force reaches about 4000 kg, an important stage is reached in the process of stressing the piece of metal. This stage is known as the 'elastic limit' and indicates the maximum dead loading to which the steel can be subjected without it deforming permanently. If the load were increased to over 4000 kg and again removed, the steel would not return to its original length but would remain permanently stretched. So, if the elastic limit was reached at 4000 kg, and the cross-sectional area was 150 mm², the stress was therefore about 27 kg force per mm² (265 N/mm²) at the elastic limit. In the elastic range, the metal, when stressed in tension, stretches by only a very small amount, but after that an increase of stress causes visible extension which can be measured easily. Above 8000 kg the steel continues to stretch and, as the load further increases, a neck gradually becomes apparent at the centre, until finally the test-piece snaps at the necked portion, leaving two pieces as shown in *Fig. 30*.

The original cross-sectional area of the test-piece was 150 mm², and this area is used in the final calculation, which shows that the steel broke at about 55 kg force per mm². Converting now to newtons as shown on page 98, the tensile strength was 55 × 9.807, or about 540 N/mm². Had

Fig. 30. Fractured tensile test-piece

the same test been done in the 'pre-SI-unit' era, its tensile strength would have been stated as about 35 tons per in².

The two broken halves of the test-piece are then fitted together, and the distance is measured between the two marks, which were originally 76 mm apart. Owing to the stretching of the metal before the break, these marks are now found to be 95 mm apart, representing an elongation of 25 per cent. This figure is recorded as a useful guide to the ductility of the steel.

In the test which has just been described, the stressing and eventual breaking of the steel proceeded in two stages: first there was an *elastic* range in which the metal did not distort permanently under stress, then there followed a *plastic* range at higher stresses, in which the metal underwent permanent distortion. Three properties have been determined: the elastic limit, the tensile strength and the elongation; this information can be put to practical use by an engineer or designer, as three simple illustrations may show.

1 *Elastic limit.* In designing a structure such as a bridge it is essential to know the elastic limit of the material, for if a girder were subjected to stress above its elastic limit it would suffer a permanent dimensional change; this might increase stresses dangerously in other girders connected to it, and lead perhaps to collapse of the whole structure.

2 *Tensile strength.* A ship's hawser might be subjected to an unusually severe stress, for example when towing a crippled ship in a gale; the tensile strength indicates the greatest stress that could be applied without the metal breaking.

3 *Elongation.* If a metal is intended to be shaped by a deep drawing operation, it must be ductile; in other words, its elongation should be high (the figure given for steel we discussed above would be counted as moderately good). A metal with only a low elongation would crack when subjected to deep drawing, and would be unsuitable for making, say, the body of a fire extinguisher.

OTHER MECHANICAL TESTS

Particularly in the testing of steels, it has become customary to measure one other figure. The 'proof stress' is the amount of stress required to cause a permanent stretch; for example, 0.07 mm on a 70 mm gauge length would be known as the 0.1 per cent proof stress. This is equivalent to the elastic limit in some metals, and is easier to measure. There has been a tendency to measure the 0.2 per cent proof stress instead of 0.1, and eventually the 0.2 per cent proof stress may be specified universally. However, any proof stress figures quoted in this book are 0.1 per cent.

Hardness is usually determined during the testing of metal products. Although it is such a familiar word, the meaning of hardness is difficult to define with precision, but technically it means resistance to deformation. This is the basis of the usual hardness tests, in which a hardened steel ball or a diamond point is pressed into the prepared surface of the metal for a given time and under a given load. If the metal is soft, a large indentation is made; if it is hard, the impression is small. The impression is measured and the hardness figure calculated on the basis of the load per unit area of the indentation.

In 1900, Johann August Brinell, of Sweden, devised a method of hardness testing that became widespread. Brinell hardness numbers are still often used in defining the hardness of metals and alloys. More recent designs of hardness testers, based on similar principles, are the US Rockwell machine and the British Vickers diamond-pyramid hardness tester. In both machines a diamond, which does not deform under heavy loads, makes the indentation. Some manufacturing companies use all three: Rockwell in production departments, Vickers in the laboratory and Brinell for testing the hardness of bulky components such as castings. Several hardness testers have automatic indentation measurements, making them suitable for mass-production control.

The Brinell hardness of some well-known metals and alloys, together with their tensile strength, proof strength and elongation, are given in *Table 9* on page 112. Various machines use different scales of hardness, but these can be correlated by the use of conversion tables. The diamond-pyramid hardness values of copper–zinc alloys are given in *Table 5* on page 61; these can be converted to Brinell by a reduction of approximately 10 per cent.

Under conditions of very sudden loading or shock, some metals behave differently from what one would expect on the basis of the tensile test; for example, some steels possess a high tensile strength but are weak under impact. In the Charpy impact test, a notched metal test-piece is broken by a swinging pendulum and the amount of energy required to break it is measured. This test can be used with either hot or cold metals.

NON-DESTRUCTIVE TESTING

In most methods of mechanical testing, part of the metal sample is fractured or damaged; one can never be fully certain that the test-piece is in every way typical of the bulk of the metal to which it relates. However, many millions of tests have shown that in most cases the test-piece does reflect the average properties of the batch of metal being processed, and hence indicates its subsequent behaviour in service.

For high-duty uses, such as in aircraft or for nuclear power, it is important that *all* metal going into service is given a thorough and complete inspection; non-destructive testing then becomes essential. During the last fifty years, advances in the development of reliable scientific instruments have made possible the 100 per cent inspection of metal parts without damaging them. X-ray apparatus is used for 'shadow' examination of castings and wrought metal shapes, provided they are not too dense or thick in section. For production where safety is of paramount importance, the X-ray apparatus is connected to video cameras so that a record of the test can be retained.

Cracks are detected by immersing the part in penetrating fluid of the paraffin type. After withdrawing the component from the liquid the outside surface is dried and time allowed for any liquid which has penetrated the crack to seep out again and be revealed on the surface. Improvements in this procedure have been made by putting fluorescent chemicals in the fluid and examining the part under ultraviolet light, or covering the surface of a casting with a lime wash or chalk to reveal the seeping fluid.

With magnetic materials such as iron and steel, a range of non-destructive tests based on their magnetic properties are used. One involves putting the part in a strong magnetic field and dusting it with fine particles of magnetic iron oxide. The iron oxide congregates where any intensive changes in magnetic intensity occur, in the region of cracks and, to a lesser extent, where sub-surface porosity exists.

With non-magnetic metals such as copper and aluminium alloys, eddy-current tests may be used to detect surface flaws. In this procedure, small localized electrical high-frequency currents are generated just beneath the surface of the tube or bar of metal under test. Search coils of copper wire pick up the residual eddy currents and compare them with eddy currents in a standard uniform specimen. This technique can reveal whether metal tubes are free from internal defects; it is particularly difficult for the human eye to detect such faults.

The workshop test of striking a metal object with a hammer and listening to the sound, characterized by the old-time railway wheel-tapper, has developed into examination by ultrasonic sound waves. These are in the same frequency range as those used for submarine detection, but they are generated in contact with the metal surface under examination and are then transmitted through the body of the metal, reflected on the far side, and picked up again by a search crystal close to the generating crystal. The search crystal transmits a signal to a cathode ray oscillograph which reveals 'echoes' if cracks, porosity or other faults are present within the body of the metal. This technique is used in the aircraft industry for the examination of metal parts before they go into service, for example spar booms in aircraft wings. Die blocks for plastic injection moulding, forging and diecasting are ultrasonically tested to determine whether the steel is free from internal flaws, thus preventing the waste of time and resources that would occur if a defect in the die block were not revealed till the final machining operation.

Over the last decade, digital instruments have revolutionized the field of flaw detection. The equipment is portable (a typical detector weighs about 5 kg), and the keypads are colour-coded with important keys strategically located for one-hand operation. They are robust and give service in a multitude of situations from North Sea oil platforms to Japanese car assembly lines.

ACOUSTIC EMISSION

Noises inaudible to the human ear can be recorded as wave-forms by suitable instruments, so we can listen to the agony of metals as they undergo stresses and strains. Such methods can be used to detect and locate deformation in metallic and non-metallic structures long before failure due to cracking would occur. In the future, acoustic emission will be developed so that, for example, the conditions in a pressure vessel could be monitored to predict the time of ultimate failure and withdraw it from service long before an accident.

FATIGUE FAILURE

The tests described so far indicate how samples of metal behave when they are stressed *once only*, whereas in service metals often have to undergo thousands, sometimes many millions of reversals of stress. For example, the shaft of a gas turbine aero engine may be revolving 16 000 times a minute, and a journey of five hours' duration would mean that the stresses in the metal alternate or fluctuate over four million times. The turbine blades themselves also vibrate in a complex manner; frequencies of up to a million per minute have been recorded, although the stresses involved were small.

The tensile test does not necessarily assess the capacity of a metal to stand up to repetitions of stresses. This was realized over a hundred years ago, when bridges of wrought iron and, later, steel were replacing stone and brick bridges. It became essential to learn about the capacity of metals to undergo many repetitions of stress, and in the early 1860s, at the request of the British Board of Trade, William Fairbairn carried out some tests in which a load was raised and lowered onto a large wrought-iron girder. It was calculated that the application of a single load of 12 tons would be required to break the girder, but Fairbairn found that if a load of a little more than 3 tons were applied 3 million times the girder would break. He concluded, however, that there was a certain maximum load, under 3 tons, which could be applied an indefinite number of times without fracture occurring. The German engineer August Wöhler carried the work further, and since that time the 'fatigue' of metals has been studied intensively.

The method developed by Wöhler for testing the fatigue of metals is used extensively. A specially shaped test-piece is gripped at one end, while at the other end of the specimen a ball race is fitted, from which a load is suspended. The test-piece is then very rapidly rotated from the gripped end by an electric motor so that under the action of the overhung load it is alternately stressed in tension and compression once each revolution. A mechanism for counting the number of revolutions or reversals of stress, often amounting to 50 million or more, is attached to the machine, which is kept running until the specimen breaks, perhaps after a short time. The load and the number of stress reversals are both noted; the same procedure is then carried out with a similar test-piece of the same metal, but with a smaller load. This time the number of reversals of stress necessary to cause failure is greater. From a knowledge of the load and the dimensions of the test-piece, the amount of applied stress is calculated; in the type of test we have been considering, a plus-or-minus sign is attached to the stress value, since there has been a measured amount of tension and an equal amount of compression. From the results of a number of such tests on steel, a graph similar to *Fig. 31* is obtained.

Notice that when the stress range is high, a comparatively small

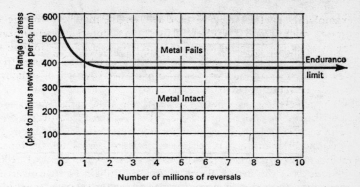

Fig. 31. The results of a fatigue test

number of alternations of stress are necessary to break the steel, but as the range of stress becomes lower a much greater number of reversals can be endured until, when the stress alternates from 385 newtons per mm² compression to the same amount of tension (the stress alternates about a mean of zero), the graph has flattened out, indicating that this range of stress could be applied an indefinite number of times.

Such a steel is said to have an 'endurance limit' of plus or minus 385 newtons per square millimetre ($\pm$ 385 N/mm²). If a slightly higher stress of, say, $\pm$ 400 N/mm² were applied, the life of that steel specimen would be only one million reversals; parts of a motor car may be subject to more than that number of alternations of stress within a year of normal use. In the case of a bridge, however, heavy stresses may be applied only a few times daily – for example, every time a train passes over, although there would be many millions of smaller stresses due to wind changes; such a bridge could be designed on the basis of a few hundred thousand applications of the maximum stress.

Many non-ferrous metals have a given life for a given stress range. Therefore, instead of the horizontal line shown in *Fig. 31*, the right-hand part of the curve would continue to slope gradually downwards, though of course for moderate stresses many billions of reversals would be needed to cause fatigue failure. *Table 8* gives the number of reversals required by an aluminium alloy for increasing ranges of stress.

It is possible to establish what is a safe load for aluminium and other non-ferrous alloys, based on a knowledge of what a part has to do in service, particularly if its anticipated length of service will elapse before the load under which it works would cause failure. Because of this,

Table 8. The endurance of an aluminium alloy under fatigue at increasing ranges of stress

Stress range (newtons per mm²)	Millions of reversals endured before failure
± 118	100
± 123	50
± 150	10
± 160	5
± 190	1

connecting rods in racing motorcyles or cars are allowed a working life of only a few hours, so that the number of reversals at the high stress range will be less than the endurance limit. This finite life of light alloy components accounts for the regular overhauls that are made to aircraft after a given number of flying hours, when some parts are renewed, partly because of wear, distortion or corrosion, but also so that they will not be stressed a sufficient number of times to make them liable to fatigue failure.

Sometimes metal parts which are subjected to fatigue stresses in service fail prematurely under fewer repetitions of stress than has been established as safe by experiment. The failure around the window frames of the cabin of the early Comet jet aircraft demonstrates the difficulty of avoiding design details which may contribute to low fatigue strength in the structure, and lead to tragic consequences. In a similar manner, scratches, dents or even inspectors' stamp marks may cause early failure under alternating stresses.

Another aspect of fatigue failure concerns the effect of corrosion on the metal's endurance. This combination of corrosion and fatigue may lower the endurance limit of a metal. At high temperatures, specific to each metal and alloy, there is no definite fatigue limit.

CREEP

When metals are used at high temperatures under uninterrupted stresses, as for example components of furnaces, they yield very slowly, so that over a period of months or years they stretch and may eventually fracture; this effect is referred to as creep. Creep at high temperatures is troublesome to the engineer, but it is not always just a high temperature phenomenon. Creep can take place at low temperatures in metals of low melting point. Thus lead sheets for church roofing gradually thicken near the eaves under their own weight.

When stress is applied to a metal under creep conditions, the following stages occur:

1 an initial instantaneous small strain, called micro-creep;
2 the primary stage of creep, during which strain or flow occurs at a decelerating rate;
3 a prolonged period, called the secondary stage of creep, during which further deformation is small and steady;
4 the creep rate accelerates, and the metal elongates rapidly and ultimately fractures. This is known as the tertiary stage of creep.

The best way of determining the suitability of a metal for service at high temperature is by means of creep tests. A specimen similar to a tensile test-piece is subjected to a constant load, surrounded by a furnace to maintain the test-piece at a constant temperature. Accurate measurements are made of the increase of length, over periods of time which may be as long as one to ten years.

Metals and alloys are not always required to endure prolonged exposure to high temperatures for years on end. Many aero engine components endure high temperatures for only 1000 to 5000 hours. This aspect has led to the development of the stress rupture test, to indicate the stress that will cause an extension at one-millionth of a centimetre per centimetre of test-piece per hour when exposed to a constant temperature for periods of, say, 100, 300, 600 and 1000 hours.

By plotting these stresses, the type of curve shown in *Fig. 32* is obtained. This indicates the breaking stress, for different temperatures, which will cause rupture after 1000 hours. Aluminium alloys, the curve for which is to the left of the figure, become weak at quite a low temperature. The next curve represents titanium alloys: in the temperature range of about 540–650°C the titanium has a creep rupture strength similar to that of steel, yet its density is only half that of steel. The curves to the right of the diagram show stainless steel, some special alloys and metal–ceramic combinations ('cermets') used in jet engines.

HOW METALS FAIL

A study of the mechanical testing of metals shows that their behaviour is not always apparently consistent, raising a number of questions about how they fail. For example, what leads to the difference in behaviour of metals when stressed (a) within the elastic limit, and (b) above it? Does the fracture of metals occur by breakage along the grain boundaries, or

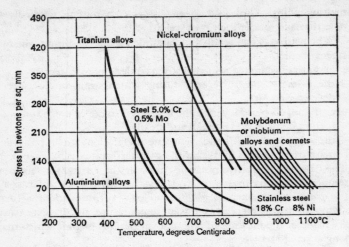

Fig. 32. Stress versus temperature for rupture in 1000 hours

are grains torn in half? In dealing with these questions, much has been learnt from studying metals under the microscope during and after stressing and by X-ray examination of the minute distortion of their atomic structures.

If a polished and etched metal such as aluminium or copper is examined under the microscope while being stressed at room temperature, a widespread change appears when the elastic limit is passed. The surface of the metal can be seen to have roughened, and each grain is marked with a number of fine parallel lines, the directions of which vary from grain to grain (*Fig. 33b*). In some of the grains two or three series of these lines may have developed. They are called 'slip bands', and are actually steps produced on the surface; they are an outward sign of the permanent deformation of the metal grains.

The mechanism of the formation of slip bands is shown in *Fig. 34*. Imagine part of a grain of metal to be stressed as indicated by the arrows in the upper drawings. At first the grain can distort elastically, and if the stress is removed it will revert to its original form as shown in the lower drawing on the left. But above a certain stress, a part of the grain slips like a pile of coins being pushed over slightly, and when this occurs part of the deformation is permanent, as shown in the bottom right-hand drawing. The planes at which the slip occurs can be observed under the microscope in the form of slip bands, as shown in *Fig. 33b* and *Plate 32c*.

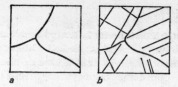

Fig. 33. Slip bands on surface of stressed metal

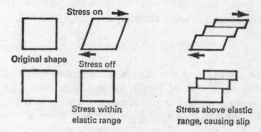

Fig. 34. The mechanism of slip

As the metal is stressed, the number and the intensity of slip bands increase until, with sufficient distortion, the structure is so confused that the original boundaries of the grains almost lose their identity. When this state of affairs is reached the metal commences to distort in another manner, by general plastic movement.

In the early stages of deformation, slip results in the hardening of the metal. It seems that the slip of one block of the grain over another results in the formation of minute fragments or 'crystallites', and it is doubtful whether slip occurs again exactly in that place because these crystallites have locally strengthened and hardened the metal. The arrangement of the atoms in a grain of metal is such that there are certain planes along which slip can take place more easily than in other directions.

The research that has been done on dislocations, previously discussed on page 60, has thrown a great deal of light on the reasons why metals are so strong, and how and why they fail under increased stress. The inspiration which led to metals being strengthened by minute fibres of other materials was derived from the spread of knowledge of the effect of dislocations, and the realization that such added fibres can produce increased strength. This will be discussed on page 285.

THE BREAKAGE OF HOT METALS

We saw on page 47 that when a stressed metal is heated above a certain temperature, it recrystallizes. When the metal is stressed while it is above this temperature, slip bands are not observed, but a type of plastic flow (possibly different from that at room temperature) occurs from the start and often results in a considerable deformation of the metal, which usually fails by cracking around the grain boundaries. The major difference between the rupture of metals when hot and when cold is that in the former state the grain boundaries are the weakest parts, while in the latter the grains themselves are the weakest.

FRACTURE MECHANICS

Metallurgists and engineers have always paid great attention to the examination of the behaviour of metals in mechanical structures, and much has been learnt from the study of components which have failed in service. In one technique a specimen is taken from a fatigue test and pre-cracked to a known amount. The test is then continued until failure occurs; in this way the effect of stresses on the propagation of a crack can be studied. Examination of the break is then carried out with a scanning electron microscope, or one of the other instruments described on page 54. Such testing has become more necessary because safety factors have often been reduced in order to lower the cost of manufacture (and sometimes to abbreviate the service lives of domestic equipment).

QUALITY ASSURANCE

The need for safety and reliability in manufactured products has led to the use of inspection routines which were first applied in the aircraft and chemical industries, but which are now widely accepted. Besides working to specifications, the inspection organization and production departments are put on their honour to follow the inspection procedures and to record and mark each batch of production so that it can be identified. The ensuring of integrity in manufacture has been developed by Rolls-Royce and by many other British concerns, as well as those in the USA, Japan and other countries. Quality control groups are formed among management and workers to assure the reliability of all products going through the plant. Statistical techniques, instrumentation and sets of gauges are used; spot checks are made on current production. The remarkable freedom from mechanical failure in aircraft provides an example of what quality assurance can achieve. For military equipment, the UK, the USA, Canada and Australia use identical quality assurance procedures. Often staff members, faced with new requirements of quality assurance, grumble about 'more red tape', but they usually agree that objectives have been achieved in a cost-effective manner, and that the status of the company has been enhanced.

CHOOSING THE RIGHT METAL FOR THE RIGHT JOB

Table 9 gives some of the mechanical properties of well-known metals and alloys. The strength has been expressed in Imperial as well as SI units. In selecting a suitable alloy for a given purpose, designers and metallurgists should work in close cooperation, bearing in mind that a wide choice of metals and alloys of varying strengths is available. Lead has a tensile strength of only 15 newtons per mm^2, while heat treated alloy steels are nearly a hundred times stronger.

The strength of a part increases in proportion to its effective section, but this way of increasing strength is not always desirable, as it increases the dead weight. Even in a bridge, the greater part of the strength is used in supporting its own weight; increase in section augments the load which the bridge has to support. The correct procedure in designing any component is to arrange for the metal to be thick at just those places where strength is required; this is why girders used in building construction are flanged, with more metal at the top and bottom, where the stresses are greatest.

A metal which is ideal for enduring one kind of stress may be unsuitable for another: cast iron, for example, is strong in compression but weak in tension. Thus it is essential to determine the type and size of the stresses to be met, and then select the material most able to meet that stress or combination of stresses. A railway coupling needs to be made in a strong but ductile metal (in other words, one with a high elongation), so that any sudden overload can be absorbed without fracture occurring; a brittle metal would not be suitable, however strong, since the smallest stretching due to severe overload, such as might arise in, for example, shunting operations, would cause immediate breakage.

Not only must the metal be fit to withstand the stresses applied in service, but it must be capable of being shaped into its finished form without difficulty. So the task of the manufacturer must be envisaged, including casting, forging, pressing, machining and heat treatment. Other special considerations of service may have to be considered. The metal may have to work in corrosive conditions, or may require some specific magnetic or electrical properties.

Table 9. Some mechanical properties of well-known metals and alloys (these figures are approximate)

Metal or alloy	Condition	As used for	Brinell hardness number	Tensile strength (tons per in²)	Tensile strength (newtons per mm²)	0.1 per cent proof stress (tons per in²)	0.1 per cent proof stress (newtons per mm²)	Elongation (per cent) on 56 mm gauge length
Aluminium	Wrought and annealed	Frying pans	27	6	92	2	31	18
Aluminium alloy with 7% magnesium	Wrought and annealed	Tubes and sheet for aircraft	80	20	309	8	124	17
Magnesium alloy with 8% aluminium	Cast and heat treated	Aircraft landing wheels	60	17	264	5	77	10
Copper	Wrought	Tubes	120	23	350	6	90	4
Copper	Annealed	Tubes	40	14	210	4	60	40
Copper	Cold drawn	Copper wire	110	28	434	26	403	4
70/30 brass	Deep drawn	Cartridges	160	35	540	30	463	10
70/30 copper–nickel	Drawn into tube	Condenser tubes	170	38	587	30	463	8
Mild steel	Hot rolled	Ships' plates	130	30	463	15	232	25
Alloy steel with 3.7% nickel, 0.8% chromium, 0.2% carbon	Forged, quenched and tempered at 400°C	Camshafts	400	86	1340	77	1188	14
Cast iron	Cast	Lathe beds	200	14*	216	—	—	—

* It would seem that when lawyers speak of 'a cast-iron case', they are displaying little appreciation of the strengths of metals: 'an alloy steel case' would be more appropriate.

11

IRON AND STEEL

The Latin word for iron is *ferrum*. Iron, steel and cast iron are classed as ferrous metals, indicating that they consist largely of iron, distinguishing them from the non-ferrous metals. They are at the root of our material prosperity: without them there might have been no great liners, railways, automobiles, skyscrapers or offshore oil platforms. As an illustration of the scale of the current levels of steel manufacture, a large integrated plant using blast furnace and oxygen steelmaking can produce 10 million tonnes of steel a year.

Steels may contain up to 1.5 per cent carbon, though the most widely used grades contain only 0.1 to 0.25 per cent. Several elements besides carbon are present in steel. Some, like manganese, have a beneficial effect, while others, for example sulphur, may be harmful; steelmakers reduce the amount of such impurities as much as possible. The addition of nickel, chromium, molybdenum, tungsten and some other elements produces a wide range of alloy steels, including high speed steels and stainless steels. Cast iron contains between 5 and 10 per cent of other elements, including carbon, silicon and manganese. It is produced by remelting iron from the blast furnace and scrap, and is the cheapest of all metals.

CAST IRON AND WROUGHT IRON

Before the Industrial Revolution steel was an expensive material, produced in only small quantities for such articles as swords and springs; structural components were made of cast iron or wrought iron. The first metal bridge in Europe, at Ironbridge over the River Severn (*Plate 15*) contains nearly 400 tons of cast iron. The project was first discussed in 1775, and work began in November 1777. In the spring of 1779, preparations were made for the erection of the ironwork. The bridge was opened on New Year's Day 1781, and the total cost was about £5000.

Wrought iron, which had a high reputation for strength and reliability, was made by a hot and strenuous process known as 'puddling', carried out in small reverberatory furnaces in which pig iron* was melted and refined by the addition of iron oxide and other substances to reduce the amount of carbon, silicon and sulphur. The temperature of the furnace was sufficient to melt pig iron (about 1200°C), but as the impurities were removed and almost pure iron was obtained, with a melting point of about 1500°C, the furnace temperature was not high enough to keep the metal molten. Using a long iron bar with an oar-shaped end, the iron was 'rabbled' into white-hot spongy lumps, bigger than footballs, which were dragged from the furnace and hammered to squeeze out most of the slag that had formed during the refining process.

A historic achievement was Isambard Kingdom Brunel's use of wrought iron for the world's first ocean-going propeller-driven iron ship, the SS *Great Britain* (*Plate 16*), once said to be the most famous vessel that ever floated except for Noah's Ark. It had five watertight transverse bulkheads, a balanced rudder and a six-blade, four-ton propeller. The hull was constructed of overlapping wrought iron plates of about 2 × 0.65 metres, riveted to metal frames. It weighed 2936 tons, nearly twice the weight of any other vessel built at that time. The ship was rigged with auxiliary sails as a standby and to save coal when the winds were favourable.

After many vicissitudes, a change of ownership and alterations, the *Great Britain* settled down to 24 years of voyages between Liverpool and Melbourne, including the transport of the first all-England cricket eleven to visit Australia. When her passenger-usefulness was ended, she was converted to a cargo vessel but was damaged in a gale around Cape Horn and finally beached in the Falkland Islands.

The story of the *Great Britain*'s triumphant return to the Bristol Great Western dry dock is not within the scope of this book, but a triple coincidence must be mentioned: 19 July 1839 was the date of laying the first plates, 19 July 1843 was her launching date and 19 July 1970 the date of her final re-docking. Prince Albert had launched the *Great Britain* and Prince Philip was on board for her re-docking. Now it is possible to see the wrought iron hull. Despite exposure to sea water for over 150 years, most plates are still more than half as thick as when they were laid down.

---

* The production of early blast furnaces was run into moulds through a main channel, the 'sow', forming ingots, the 'pigs'. Nowadays the metal is converted to steel and is generally called hot iron.

Fifty years after the *Great Britain* was launched, Gustave Eiffel's famous tower was erected for the 1889 Paris Exhibition. Steel was considered as a possibility, but wrought iron was cheaper then, so the Eiffel Tower was constructed by Forges de Wendel using 7 341 000 kg of wrought iron. Eiffel, renowned as a brilliant engineer, had previously advised on the design and manufacture of the Statue of Liberty, completed in 1886. A hundred years later, when the statue was refurbished, careful testing revealed no cracks in its wrought iron; it has an estimated life of another 500 years.

EARLY STEELMAKING PROCESSES

In August 1856 Henry Bessemer announced his invention of a process which made it possible to produce steel cheaply and in large quantity. He proposed burning away the impurities by blowing air through molten pig iron, an idea that appeared fantastic and dangerous to the Victorian ironmasters. However, they invested large sums of money in his process, only to be scandalized that they could not make it work. Bessemer paid back all their money, spent several thousand pounds in discovering what had gone wrong, and proved that his own experiments had been with an iron containing only a small percentage of phosphorus; they had unfortunately tested his converter with iron of a high phosphorus content. He tried to persuade his disillusioned clients to try his process again, but they had been 'once bitten'. Bessemer then decided to go into production himself. He built his own steel works in Sheffield, and was soon making nearly a million tons of steel a year. He continued to meet opposition and once, when he attempted to interest a railway engineer in the possibility of using steel for railway lines, received the reply, 'Mr Bessemer, do you wish to see me tried for manslaughter?'

Bessemer converters, like huge concrete mixers, could be tilted to receive 25 to 50 tons of molten pig iron, and were then brought upright. Air blown through holes in the base of the converter forced its way through the molten metal. Oxygen from the air blast combined with some of the iron, producing iron oxide which reacted with the silicon and manganese to form a slag. The carbon was removed as carbon monoxide, part of which burned and formed carbon dioxide. The complete operation, from one tapping to the next, took 25 to 30 minutes. No external fuel was necessary; the heat requirement was furnished by the oxidation of the impurities, and some of the metallic iron. The converter was tilted and its metallic contents were poured into a large ladle. An alloy containing manganese was added while the metal was being poured into the ladle; the manganese combined with dissolved iron oxide, thus removing it from the steel. Then anthracite was added to bring the steel to the correct carbon content.

Twenty-three years after the introduction of the Bessemer process a tragedy occurred which drew attention to the possibilities of construction with steel. On the night of 29 December 1879, a bridge over the River Tay collapsed while a train was crossing, and 78 lives were lost. The bridge, built of wrought iron and cast iron, with 84 spans each 70 metres in length, had been described as one of the engineering wonders of the world. The failure of the structure was due not so much to low strength in the wrought iron, as to faulty manufacture of cast-iron columns and inadequate allowance for the stresses produced by wind and flood which were so great on that stormy December night. Nevertheless, the disaster stimulated engineers to reconsider what materials should be used for construction; it was realized that steel, which by that time was being produced on a considerable scale, was a more suitable engineering material for bridges than either cast iron or wrought iron.

THE OPEN-HEARTH PROCESS

In England, the German-born Charles William Siemens and his brother Frederick, and later the brothers Pierre and Emil Martin in France, were responsible for a method of making steel from molten pig iron and scrap in large reverberatory furnaces. They experimented with a method of 'heat regeneration', making the outgoing burnt furnace gases pre-heat the incoming gaseous fuel. By this means a sufficiently high temperature was obtained to treat large quantities of metal and to keep it molten throughout the process, enabling it to be cast into ingots when the refining was complete. The open-hearth furnace was so called because the molten metal lay in a comparatively shallow pool on the furnace hearth (*Fig. 35*).

Usually scrap steel was charged and heated first in the furnace and molten pig iron added to it; thus the impurities were diluted and the refining process was shorter than if the entire charge had been of pig iron. Iron oxide, in the form of iron ore or scale, was added. This, together with oxygen in the furnace gases, oxidized the impurities, the carbon being removed as carbon monoxide. Some silicon and manganese were also changed into their oxides, which reacted with added lime to form a slag. At the end of the process, when the impurities had been brought down to the required level and the steel had been tapped, additions of ferro-manganese, ferro-silicon or other alloys were made in the ladle, to bring the steel to the correct composition.

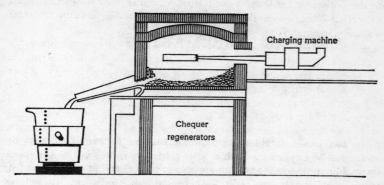

Fig. 35. Open-hearth furnace

ACID AND BASIC FURNACES

Bessemer and open hearth furnaces were lined with either of two types of refractory, depending on the composition of the blast furnace irons that were being converted to steel. Silica brick linings were known as 'acid', and such furnaces, as the Victorian ironmasters found, were unsuitable for dealing with irons containing more than a few tenths of a per cent phosphorus.

Later it was found that if converters were lined with 'basic' firebricks, containing magnesite or dolomite, highly phoshoric irons could be treated satisfactorily. Lime was added to the bath to combine with phosphorus and silicon, forming a slag containing calcium phosphate and calcium silicate. The basic lining of the converter provided conditions under which the added lime could react with the impurities without destroying the refractory.

Open hearth furnaces were also built with either acid or basic linings. In Britain the last acid open hearth went out of commission in 1975, and the last basic one in 1965.

The complete cycle of operation lasted from five to fourteen hours, and the open-hearth furnace of average capacity dealt with 10 to 25 tonnes an hour. Remembering from Chapter 3 that a modern blast furnace produces 5000 tonnes of iron per day, and that some of the newest ones make twice that figure, it will be seen that one blast furnace could keep at least eight open-hearth furnaces in operation. This rather inefficient performance – compared with the enormous output of the modern blast furnace – justified the emergence of the highly productive oxygen process for steel.

12

MODERN STEELMAKING

In the late 1950s, steelmaking took a great leap forward thanks to the production of oxygen on such a scale that it is measured by the tonne (about 740 cubic metres at atmospheric pressure) and at a fraction of its former cost. Plants are built next to modern steelworks, capable of providing several hundred tonnes of oxygen a day.

The first development began in Austria, whose steelmakers had depended to a large extent on Bessemer 'acid' converters and who gained confidence by enriching the air blast with oxygen, steam and oxygen, or carbon dioxide and oxygen. The local Austrian ore was too low in phosphorus to enable the air-blown 'basic' Bessemer method to be used, while the amount of steel scrap available was not high enough to make open-hearth production economic. Thus a combination of circumstances in Austria provided the incentive for this major development, which radically altered the steel industry all over the world.

The Austrian process was called LD, an abbreviation of *Linz Düsenverfahren*, the Linz lance process. (It is often suggested that the name was derived instead from the initials of two separate plants, at Linz and Donawitz.) The LD process consists of blowing a jet of almost pure oxygen at high pressure and at supersonic speed onto the surface of molten iron from the blast furnace; it takes place in a converter, as illustrated in *Fig. 36*. The vessel remains stationary in a vertical position throughout the blow. The speed of the reaction made high speed analytical control essential, with the installation of automatic computerized 'press button' analysis machines. The development of oxygen steelmaking caused a tremendous increase in productivity: in the 'pre-oxygen' era it took ten hours to make 250 tonnes of steel in an open-hearth furnace; now it can be done in less than half an hour.

The original LD process was limited to the treatment of iron whose phosphorus content was under 0.3 per cent. Then French and Luxembourg steelmakers discovered that the injection of lime with the oxygen allowed them to treat blast furnace iron of much higher phosphorus

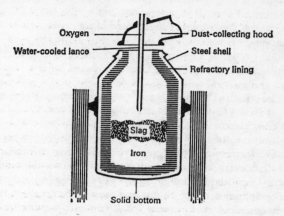

Fig. 36. An LD top-blown converter

contents – up to nearly 2 per cent. This version of the process is known as the BOP (basic oxygen process).

At present the world steelmaking community refines molten metal from the blast furnace, plus scrap, by blowing oxygen into the vessel and onto the bath in one of three ways, each with a number of varieties: from the top, from the bottom or combined from both directions. The first group includes LD and BOP, described above, the second includes the German OMB (Oxygen-Maximilianhütte-Boden) and the American QBOP (the Q standing for quiet, because it is less noisy and violent than the basic oxygen process). There are at least fifteen combined blowing processes, in Europe, the USA and Japan. One of these, known as BAP (bath agitated process), is a technique developed jointly by British Steel and Royal Netherlands Steel. In normal practice interaction between the oxygen jet and the surface of the metal results in the formation of slag which is too highly oxidized. By stirring with an inert gas injected through the bottom of the furnace, the composition of the steel is improved and the operation made more controllable. The oxygen is injected from the top of the converter and the stirring action is provided by argon and nitrogen blown from the bottom. The thickness of the refractory lining had to be increased from 600 to 800 millimetres, but the introduction of the bath agitated process required very few other modifications of the furnaces.

New methods of oxygen blowing have been developed; for example, in one system oxygen is blown through annular tuyères in the bottom of the vessel. Computerized control of operations lies at the heart of modern oxygen processes. A water-cooled lance carrying a combined analytical and thermostat probe is lowered into the bath nine-tenths of the way into the blowing time to analyse the carbon content and register the temperature of the steel. Data thus collected are compared with a computer-based 'model', and near the end of the blow the computer generates any changes necessary for the final stages of the programme. Every part of the process is analysed to achieve maximum saving of energy and precise control of the conditions.

The price and availability of steel scrap vary quite widely, and steelmakers have introduced techniques for calculating the amounts of various grades of scrap to be charged. In some plants the rapid and efficient output from bigger and better blast furnaces has reduced the proportion of scrap that needs to be put into the oxygen furnace.

Oxygen processes present special problems of moving immense amounts of hot iron scrap and liquid steel to and from the relatively small space around the furnace. Compared with its large output, there is little room or need for human operators! Fast production, fast measurements and fast tapping are essential. A single oxygen furnace can produce over 3 million tonnes of steel per year, which means moving 6 million tonnes of metal in and out.

ELECTRIC ARC STEELMAKING

Open-hearth furnaces could handle the melting and treatment of large amounts of scrap but, by the time of their demise, oxygen furnaces were more suitable for taking hot metal from the blast furnace together with comparatively small amounts of steel scrap. Electric arc furnaces were at first intended to produce high grade steel, but now they also handle bulk production of 'ordinary' steel, using large amounts of scrap. About 200 million tonnes of steel are now produced worldwide with electric arc furnaces, representing about a quarter of all steel production.

The basic design of a typical furnace is shown in *Fig. 37*. It has a cover which can be swung to one side to enable scrap and other materials to be charged. The cover has four holes, three at the centre, through which the electrodes are positioned and a fourth for exhaust-gas extraction. The furnace hearth is of magnesite bricks, onto which a layer of dolomite is rammed. The lower walls of the furnace are lined with refractory, while the upper walls and the roof, which are not in contact with molten metal,

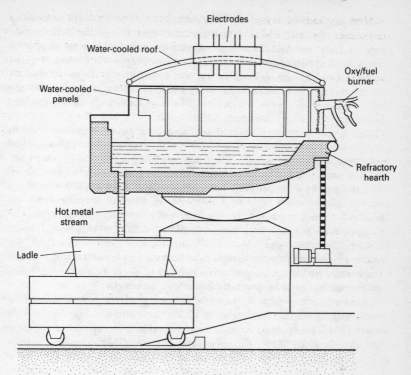

Fig. 37. Electric arc furnace

are water-cooled. Most electric furnaces range in size from 3 to 7 metres
in diameter, though there are larger furnaces in operation, including one
in the USA 11.4 metres in diameter. Typical furnaces have capacities of
about 120 tonnes per heat.

Scrap is collected from stockpiles by a crane equipped with a magnetic
hoist. The furnace electrodes are raised, and the cover swung clear. A
large container, which can be opened at the bottom by the crane driver,
is lowered and delivers its contents to the furnace. Lime is added to
form a slag, along with carbonaceous material. The furnace is closed and
the electrodes lowered onto the scrap. As soon as the electric circuit is
made the electrodes are raised slightly, causing an arc to be struck

between the electrode tips and the steel scrap. The temperature of the arc reaches about 12 000°C. When the first scrap has melted, one or two more charges are added. When all the metal is melted, a sample is analysed, the electrodes are raised, the furnace is tilted slightly, and oxygen is injected to oxidize the carbon down to the required amount and to reduce the silicon and manganese contents, followed by the addition of a slag to reduce sulphur. The furnace is fully tilted, the slag removed and the steel is tapped.

During the last thirty years there have been many improvements in the operation and efficiency of electric arc furnaces. In 1965 a typical cycle time was three hours, 630 kWh of electric power was required per tonne of steel and 6.5 kg of electrode material was consumed per tonne. Now the power consumption has been reduced to 400 kWh per tonne and the electrodes to 2.15 kg per tonne. A modern furnace produces about 100 tonnes per hour. The high cost of electricity has led to the use of supplementary oxy-fuel burners, directed at the three cold spots between the electrodes. Also, bath stirring has been introduced. To replace alternating current several pilot furnaces have been commissioned to operate with direct current; this has led to some further economies, for example in the consumption of electrode material.

There are about fifty different grades of steel scrap available in an ever-changing market, and what is known as computer-based least cost mix (LCM) is employed to match the selection of scrap to the required steel specification. While the speed and efficiency of the operation have been increased, so have the demands for high quality, leading to a greater dependence on computer control of every aspect of electric arc steel production. Like most industrial processes it has been necessary to improve environmental control, especially for the treatment of fume emission.

LADLE METALLURGY

More and more stringent demands are being imposed on steelmakers by their customers: stricter and closer control of composition, cleaner steels in which inclusions of slag and oxide are removed or made innocuous, and all this at the lowest price. These requirements have led to many 'new looks' in steel manufacture; one important development involves metallurgical treatments being performed in a large ladle, into which the molten steel has been poured. A director of one steel plant summed up the process as 'letting the melting furnace do what it is supposed to do – make a master steel – then making the adjustment in the ladle'.

The range of ladle treatments includes refining, lowering the sulphur content, control of inclusions, temperature control and alloying. There are several competitive techniques available. In one process an inert gas such as argon is injected to remove the harmful impurity sulphur, followed by the insertion of calcium wire to remove inclusions. In another method, known as wire feed injection, powdered reagents encased in a metal sheath or supplied directly as a wire product are added. The wire is coiled on a drum stationed on a platform overlooking the ladle and is fed down into the molten steel. Some wire injectors provide calcium–silicon additions, while others are used to inject alloying elements. Most ladle treatments also involve stirring the steel, usually by an inert gas injected through the bottom of the ladle. Deoxidants are added, the most commonly used, in ascending order of 'power', being manganese, silicon and aluminium. Vacuum degassing in the ladle was first developed to remove hydrogen, but it was also found to give cleaner steels, and nowadays this is insisted upon by users.

CONTINUOUS CASTING OF STEEL

A major development in which European steelmakers played a large part has been the production of sections direct from liquid metal. Continuous casting of steel was introduced in the mid-1950s and has grown rapidly: the present annual world capacity is about 500 million tonnes, representing about two-thirds the total tonnage of steel; it is expected to reach 90 per cent by the end of the century.

The continuous casting of steel was developed from the similar process for non-ferrous metals, described on page 77, but early attempts to cast steel in this way were beset with difficulties, the most troublesome of which was the sticking of the solidified steel shell to the walls of the mould. The use of a lubricant and the introduction of a reciprocating movement of the mould were two of the many developments which helped to overcome the problems, and enabled continuous casting to usher in a major new era in the steel industry.

In the original continuous casting machines for steel, molten metal was poured from a height of over 13 metres above ground level, with the cast section of metal descending vertically and being progressively water cooled. The operating height was then reduced by bending the section after solidification; a further reduction in height was achieved by casting the section in a curved mould and cooling on the curve instead of vertically. *Figure 38* shows the details of such a modern continuous casting machine. An eight-strand machine illustrated in *Plate 17*,

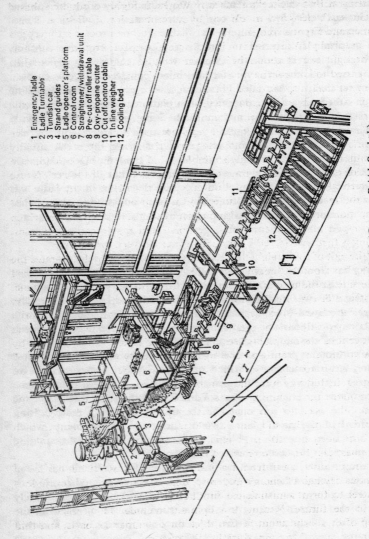

1 Emergency ladle
2 Ladle Turntable
3 Tundish car
4 Strand guides
5 Ladle operator's platform
6 Control room
7 Straightener/withdrawal unit
8 Pre-cut off roller table
9 Oxy-propane cutter
10 Cut off control cabin
11 In line weigher
12 Cooling bed

Fig. 38. A single-strand continuous casting plant for steel slab (*Courtesy British Steel plc*)

operating at British Steel's Lackenby Works, is highly productive, based on 'Concast', designed in Europe by a combination of British, Swiss, German and French companies.

In a typical plant, steel is run into a ladle, transported to the casting bay and brought to rest above the Concast machine. A stopper rod or slide gate is used to control the flow of steel into a container called a tundish. Once a set steel level is attained in the tundish, nozzles are opened and the molten steel feeds into the open-ended mould set in a frame with internal water cooling. At start-up, a dummy bar constructed of chain links is fitted into the mould bottom and rests between the arc formed by the machine roller guides. The molten steel in the mould meets the dummy bar and freezes to it. At a certain level of steel in the mould, two operations are immediately initiated: the mould unit begins to reciprocate in a vertical direction, and the dummy bar is withdrawn through and down the machine. As the solidifying cast slab passes from the vertical to the horizontal plane it is supported by a series of rollers which are grouped into segments for operational and maintenance purposes. Cooling is further effected by spraying water between the rollers.

At the exit of the straightener unit a roll is lowered which separates the dummy bar from the length of cast steel. The dummy bar is either lifted to one side of the machine or hoisted onto the ramp above the machine. The strand of cast metal continues travelling along the discharge rollers to a gas torch cut-off station. At predetermined dimensions this torch will automatically cut the strand to required lengths. The end products of continuous casting plants are square-section 'billets' up to 150 mm by 150 mm, 'blooms' from that size up to 400 mm by 600 mm, and 'slabs' which are up to 250 mm thick by 2725 mm wide. Other shapes are produced, including round and octagonal sections.

By suitable programming of the metal supply, 'sequence casting' now obviates the need to run out the cast steel after each ladle has been emptied. In one British plant 183 000 tonnes of slab were cast in a 28 day continuous operation. A Japanese company, using ladles holding 160 tonnes, cast 176 000 tonnes in 43 days.

The rapid move away from traditional ingot casting of steel has led to enormous savings of energy and cost. The direct rolling of continuous cast steel to form semi-finished products will certainly become widely used in the future. Recent developments include the near-net shape casting of thin slabs about 50 mm thick on a commercial basis, and thin strip in the range 2 to 6 mm thick in pilot plants.

STEEL MINIMILLS

Minimills get their name, not because they are small, but because their operation does not include the production of hot iron. The basic design of a typical plant includes one or two electric arc furnaces, either for melting scrap or taking iron sponge or pellets from a direct reduction plant. The molten steel goes to a continuous casting machine and the billet which emerges is cut to convenient lengths, which go to a rolling mill. The products of minimills include wire rods, reinforcing rods for concrete, bar products, light I-shaped beams and channels. High productivity is characteristic of the more profitable minimills; rolling speeds have more than doubled during the past decade.

These plants can be located in areas where cheap fuel is available and where there is ample iron and steel scrap. One example is in Texas, about 1500 miles from the big USA steelworks, but in an area with a great deal of manufacturing industry. It would be expensive to obtain steel from Pittsburg and then return scrap the same long distance, so the minimills provide the material and deal with the available Texan scrap. At present minimills in the USA produce about a quarter of their total steel output. Britain does not have the immense distances of America, but it has some areas in which the operation of minimills is logical, for example close to the reinforced concrete and scrap metal markets near London.

Minimills have the advantage that they are designed to make comparatively simple, locally required products. The capital cost per tonne of steel produced is only half that for a conventional steelworks. They can be profitable in the right circumstances, though they need alert managements, capable of making rapid decisions. Such managements have been quick to adopt new methods, including electromagnetic stirring and ultra-high-power arc furnaces. Their dependence on the technology of continuous casting has been crucial to their success, but they can be crippled economically if the price of scrap or the cost of electric power increases drastically.

COATED STEEL

One of the most significant developments during the last decade has been the growth in the production and use of steel strip with a protective metallic or coloured organic coating. Much of the impetus for this development came from demands for corrosion protection of automobiles, beginning in America and Japan and then in Europe. Nowadays new car purchasers want long guarantees for automobile bodywork. Three types of coating process are available: (a) heat treatment followed by galvanizing, (b) electroplating and (c) organic coatings.

A typical multi-layer coating has a substrate of galvanized steel covered by a pre-treatment layer such as zinc phosphate. The primer coat contains anticorrosion pigments. Finally, a wide range of coloured top coats can be applied, either as liquid paints or as a laminate secured to the strip by adhesive.

In addition to the vast automotive market, the building industry and the manufacture of domestic equipment have become major users of coated steel, and the range of colours is splendid. The coated steels are shipped as coils from which components of cars or washing machines are made.

QUALITY CONTROL

During the last twenty years every industry requiring steel has demanded and received standards of quality that would have been impossible before. In British Steel over ten thousand personnel, from senior management to operatives, have received training in statistical process control, and similar efforts have taken place in steel companies all over the world.

Pipeline steels require low levels of carbon, manganese and phosphorus. Exacting demands come from automobile manufacturers. Other industrial activities, ranging from the construction of multistorey buildings to the production of the 5 mm wire used in immense quantities for bridge building, have their own specifications. Offshore platforms require steels of maximum toughness, weldability and surface quality, and a sulphur content of less than 0.002 per cent. As an example of the vital cooperation between manufacturer and client, *Plate 18* shows the immense jacket for the Magnus Oilfield, designed to withstand the stormy waters of the East Shetland Basin. This huge structure, 85×85 metres at the base, 65×63 metres at the top and 212 metres from top to bottom, contains over 30 000 tonnes of steel: when the 36 piles were added, the total weight came to over 40 000 tonnes. The size of the structure is emphasized by the double decker bus parked alongside it.

13

THE ROLE OF CARBON IN STEEL

When carbon is alloyed with iron, the hardness and strength of the metal increase. For example, a steel containing 0.4 per cent carbon is twice as strong as pure iron, and with about 1 per cent carbon nearly three times as strong, though as the carbon content rises the ductility diminishes. Alloys of iron with 1.5 to 2.5 per cent carbon are rarely used. When it contains more than 2.5 per cent carbon the metal is classified as cast iron, characterized by good castability, moderate strength and hardness, but low ductility. The iron–carbon alloys are classified according to their carbon content, although there is no precise demarcation (*Table 10*).

Table 10. Classification of steels by carbon content

Description	Approximate carbon content (per cent)
Mild steel	up to 0.25
Medium carbon steel	0.25–0.45
High carbon steel	0.45–1.50
Cast iron	2.50–4.50

Depending on the application to which the steel is put, a suitable iron–carbon alloy is selected to provide the required properties. For example, in a suspension bridge the cables are of 0.8 per cent carbon steel, the roadway plate is 0.15 per cent, the reinforcing bars are also 0.15 per cent, and prestressed wire in the reinforced concrete is 0.8 per cent. *Figure 39* shows some of the uses of steels of various carbon contents.

The behaviour of carbon steels which have been slowly cooled is attributed to three causes (see also the discussion of the atomic structures of iron on page 58).

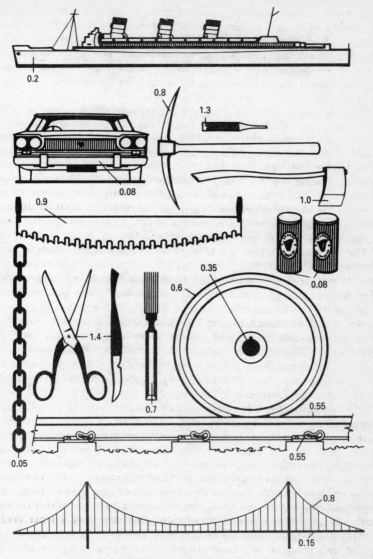

Fig. 39. Some applications of steel (figures indicate percentage of carbon)

1 At room temperature the atoms of iron are arranged in the body centred cubic lattice pattern (*Fig. 16*). When heated to 906°C, the iron atoms spontaneously rearrange themselves into the face centred cubic lattice pattern (*Fig. 15*). Why this takes place is not fully understood, but it is a fact of great importance to a metal-using civilization, for if the change did not occur, steels would not be amenable to heat treatment.

2 At room temperature less than 0.01 per cent carbon can exist in solid solution in slowly cooled iron of body centred atomic arrangement, but at high temperatures up to 1.7 per cent can be taken into solid solution in face centred cubic iron, this maximum solubility being attained at 1145°C.

3 When molten steel is allowed to solidify and cool slowly, each carbon atom unites with three iron atoms to form a compound, iron carbide, Fe_3C. Although a compound of a metal and non-metal, it is similar in behaviour to the general class of intermetallic compounds discussed on page 50, and its presence confers hardness.

The amount of carbon in the steel has a pronounced effect on the temperature at which the change in the atomic lattices occurs. Thus, if pure iron is gradually heated from room temperature to bright yellow heat, but not melted, the arrangement of the iron atoms changes from body centred to face centred cubic at 906°C. With about 0.3 per cent carbon the change commences at 730°C, and is complete at about 800°C. With 0.8 per cent carbon the change begins and completes itself at 730°C. Thus, when carbon is present the atomic reshuffling commences at 730°C and concludes at that or some higher temperature, according to the carbon content. *Figure 40* illustrates this, and the figures given above can be checked by reference to the diagram.

The V-shaped graph is reminiscent of that representing the effect of one metal on the melting point of another, where a eutectic is formed with a minimum melting point (*Fig. 8*). However, the behaviour of steel which has just been described concerns the effect of carbon on the change of atomic arrangement in *solid* iron. The structure of the 0.8 per cent carbon composition is called 'eutectoid', and 730°C (or, to be precise, 732°C) is called the eutectoid temperature. The actual eutectoid composition and temperature depend to some extent on the purity of the iron. A commercial grade of steel may show a eutectoid at 0.9 per cent and a eutectoid temperature at 700°C. For the following discussion, however, we have assumed that the iron–carbon alloy is pure.

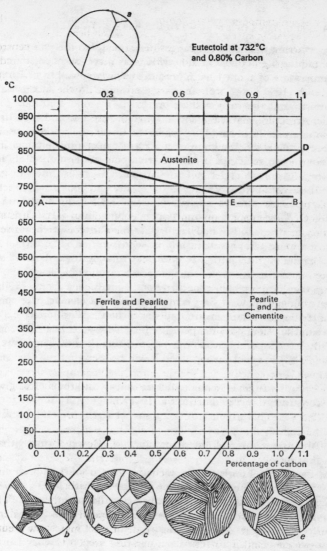

Eutectoid at 732°C and 0.80% Carbon

Fig. 40. The microstructure of steel depends on temperature and carbon content

THE STRUCTURE OF STEEL WITH 0.3 PER CENT CARBON

By referring to *Fig. 40*, it is possible to picture what happens to a steel containing 0.3 per cent carbon which has been solidified and cooled to a temperature of about 1000°C, and then slowly cooled from that temperature. At 1000°C the iron atoms are arranged in the face centred cubic pattern, and the carbon therefore exists in solid solution. The upper microstructure marked *a* shows what the steel would look like at this temperature. If it could be examined under the microscope, it would be seen to exhibit the grains characteristic of a solid solution. The name given to the structure of this iron–carbon solid solution is 'austenite', after William Chandler Roberts-Austen, an English metallurgist of the late nineteenth century.

When this steel is cooled to about 800°C, a change takes place, indicated by a point on the upper sloping line, CE. The iron atoms begin to revert to the body centred cubic pattern which normally holds only minute amounts of carbon in solid solution. The carbon atoms do not come out of solution at once, but migrate towards areas where the iron atomic lattice is still in the face centred cubic form, thus increasing the local carbon concentration. Finally, at about 730°C all the remaining regions of face centred cubic iron change over to the body centred cubic arrangement, and the carbon can no longer be contained in solid solution. This temperature is shown by the line AEB.

By referring to illustration *b*, representing the microstructure of slowly cooled 0.3 per cent carbon steel, it will be seen that the final structure is duplex.

1 About two-thirds of the structure consist of grains of iron, the white constituent in the illustration. In metallography this is identified as 'ferrite'. The presence of the grains of ferrite confers ductility on the steel.

2 About one-third of the structure is a layered formation, known as pearlite.

As the hot steel cools slowly, carbon from the solid solution is deposited as iron carbide, Fe_3C, in layers alternating with layers of the surplus ferrite; this formation is the darker constituent in illustration *b*. This layered structure is called 'pearlite' because, when viewed under the microscope under oblique lighting, it has the iridescent appearance of mother-of-pearl. This formation was first seen by Henry Clifton Sorby when he was using the metallurgical microscope in the early 1860s. A photograph of pearlite is shown in *Plate 30b*; it is the eutectoid which was formed at 732°C.

With Fe_3C (usually known as 'cementite') alone, the steel would be brittle; with ferrite alone it would be soft. Pearlite combines the good properties of both constituents, though pearlite alone would be too hard for structural uses. The most widely used steel, containing 0.2 to 0.3 per cent carbon, has a structure of one-third pearlite and two-thirds ferrite when the steel is in the slowly cooled condition.

A steel containing about 0.8 per cent carbon is completely pearlitic when it has been cooled slowly (illustration *d*). A low carbon steel has only a small amount of pearlite, while in a steel with over 0.8 per cent carbon the structure consists of pearlite plus surplus cementite (illustration *e*), and such steels possess great hardness and strength, but low ductility. By referring to *Fig. 40* the structure of other steels can be pictured; thus a slowly cooled steel with 0.6 per cent carbon is composed of about two-thirds pearlite and one-third ferrite (illustration *c*).

THE HEAT TREATMENT OF STEEL

The simple experiments described below may be enlightening. Two old-fashioned steel knitting needles, a gas ring, a bowl of water, a piece of sandpaper and a pair of pliers are needed. (Steel knitting needles are 25 cm long and are magnetic; they are not to be confused with anodized aluminium needles which are 30 cm long. If steel knitting needles cannot be found, a large darning needle or a bodkin will perform satisfactorily.)

1 Bend one needle slightly, and feel how tough and springy it is. Now hold the needle in the flame, using the pliers, with the bowl of water close at hand. When it is bright red, dip the end of the needle as quickly as possible into the water. The needle should still be red hot as it is being quenched. Then try to bend the quenched end; it is hard and brittle and will snap off.

2 Take the second needle and heat it until it is red hot; maintain it at this temperature for 10 to 15 seconds. Then withdraw it very slowly, so that it cools gradually. If you now test this end (which you have just 'annealed') it will bend, like a piece of soft wire, and furthermore it will remain bent.

3 Heat the needle which has been softened to bright red heat, and quench it rapidly to make it hard and brittle, as in Experiment 1. Clean the needle with sandpaper and hold it above the flame so that it is warmed until a straw colour develops on it; it must not reach even dull red heat. When the tint appears, take the needle away and let it cool down. If you bend the end of the needle, which you have 'tempered', you will find that it is tough and springy.

Table 11. Three heat treatments for steel

Treatment	Name of process	Resulting condition of steel
Heated to red heat (850°C) and quenched	Hardening by quenching	Hard and brittle
Heated to red heat and slowly cooled	Annealing	Soft and not springy
Heated to red heat and quenched; warmed to about 250°C; cooled in air	Hardening and tempering	Tough and springy

Thus three heat treatment processes have been performed on a domestic scale, processes which are being continually carried out, under rather closer control, in factories all over the world. The processes are summarized in *Table 11*. The striking variation of properties of steel as obtained in these experiments is associated primarily with its carbon content (in this case about 0.8 per cent), and secondly with the rate of cooling from bright red heat.

The microstructure of the slowly cooled steel of the needle is similar to that shown in *Fig. 40c*, and its structure consists of pearlite with some ferrite. The atomic lattice of the steel, when heated to bright red heat, changes to the face centred cubic arrangement, and the carbon is taken into solid solution. This would give the simple microstructure of austenite. From *Fig. 40* it will be seen that, on *slowly heating*, the change in structure in a 0.7 per cent carbon steel starts at 710°C (the line AEB) and is complete at about 730°C (the line CED). The red-hot steel of the needle, when quenched in water as in Experiment 1, cools so rapidly that the carbon atoms have no time in which to come out of solid solution to form cementite, which is one constituent of pearlite. Because of the enforced presence of these carbon atoms the iron atoms can revert only to a distorted form of the body centred cubic arrangement. This severe distortion is responsible for the great hardness and brittleness produced as a result of Experiment 1. Under the microscope, this steel in the quenched condition shows a type of structure which is illustrated in *Plate 30c* and which is called 'martensite' after Adolf Martens, a nineteenth-century German metallurgist.

Few steels are used in the extremely hard martensitic condition, and the reheating operation of tempering is carried out to reduce the brittleness of the steel while still retaining much of the hardness, as was demonstrated in Experiment 3.

If the quenched steel is tempered at a temperature as low as 250°C its hardness is reduced only slightly, but if tempering is carried out at a higher temperature the reduction of hardness is greater. Ordinary steels are not often tempered between 250 and 500°C because in that temperature range they tend to become embrittled.

When martensite is tempered by heating it to a temperature below 700°C, particles of cementite begin to form, and in the high tempering temperature range of about 500 to 700°C the globules of cementite attain sufficient size to be seen under the microscope. The structure shown in *Plate 30d* is of a steel first hardened and then tempered at 300°C; at this temperature the cementite particles have not become visible.

Two important points are to be noted in connection with the quenching and tempering of steel.

1 To get a fully hardened structure the steel should be heated to a temperature above that represented on the line CED on *Fig. 40* and then rapidly cooled. It will be seen, therefore, that the appropriate temperature before quenching depends on the carbon content of the steel.

2 Tempering does not restore the pearlitic structure. Before that can be done, the steel has to be heated to a temperature above that represented by the line CED and then slowly cooled. Tempering is carried out to relieve residual stresses in the quenched steel, to produce the ultimate properties required after treatment.

In the industrial heat treatment of steels, intermediate rates of cooling from high temperatures are also used. Normalizing consists of heating a steel to red heat (above the line CED) and cooling it in air. This gives a 'normal' ferrite–pearlite structure, similar to that shown in *Fig. 40d*.

Although water was once used as a quenching medium, proprietary mineral oils, polymers and inert gases are used today. Too severe a quench would cause large blocks of steel to crack; furthermore, many steels containing alloying elements do not require rapid quenching. Heat treatment used to be done in quite primitive and unhealthy conditions, but now computerized process control has entered the field of heat treatment management. The furnaces are sometimes linked to automated handling systems for all movement of materials.

Most case hardening and oil hardening operations are now carried out in multi-chamber, controlled atmosphere, sealed quench furnaces. Following heating in a sealed muffle furnace, the components to be treated are moved to an integral cooling chamber where quenching is carried out by lowering the steel into an oil quench bath under a controlled atmosphere. Typical components treated in these furnaces include automotive steering and suspension components, transmission gears and shafts, aircraft undercarriage parts and spring spikes for attaching railway rails to sleepers.

For some heat treatment processes, vacuum furnaces are required. *Plate 19* illustrates two of the largest vacuum furnaces, with a load capacity of over 2 tonnes and an operating temperature range from 260 to 1350°C. Their work chambers are 1.8 metres in diameter and 1.8 metres high. Parts for treatment are cooled in an inert gas (generally nitrogen, although argon is required for titanium alloys). The neutral conditions in the vacuum furnace prevent oxidation or recarburization from taking place. Vacuum furnaces are required for the heat treatment of casings and turbine blades for jet aero engines and medical prostheses, for example for replacement hip joints. Stainless steel kitchen sinks are nowadays vacuum annealed between deep drawing operations.

There have been notable advances in vacuum heat treatment in recent years. The impetus for this has come from expansion in the aircraft industry, and its requirements for high temperature treatments of complex stainless steels, titanium alloys and the nickel-based Nimonic alloys (discussed on page 191).

SURFACE HEAT TREATMENTS

Surface heat treatments are employed to improve the performance of engineering components, tools and dies by enhancing surface hardness and resistance to wear, fatigue and, in some instances, corrosion. The techniques for ferrous metals fall into four categories:
1 high temperature thermochemical treatments, such as carburizing;
2 low temperature thermochemical treatments, such as nitriding;
3 selective surface treatments, such as flame-hardening;
4 treatments involving the deposition of ultra-hard metallic compounds, such as boronizing.

HIGH TEMPERATURE TREATMENTS

When mild steel or low alloy steels are heated to temperatures above 900°C in a medium which provides a source of carbon, the carbon is absorbed at the surface of the steel by diffusion. The depth of enrichment depends on the time and temperature of the treatment. A mild steel carburized at 925°C for eight hours would receive a 'case' about a millimetre thick and, ideally, a carbon content at the surface of about

0.9 per cent. After subsequent quenching, the case-carburized steel has a hard surface layer, but retains the toughness of the low carbon steel in the core.

Until a few decades ago the carbon case was applied by immersing the steel in a bath of molten sodium cyanide, a toxic compound of sodium, carbon and nitrogen. Other salts, such as sodium or barium chloride and sodium carbonate, were added to lower the melting point of the mixture. However, environmental considerations prompted the development of systems in which cyanide was eliminated. As the old cyanide treatment furnaces became discontinued, 'fluidized bed' heat treatment plant came to replace many of them. A controlled flow of gas passes through a bed of alumina particles, creating a fluidizing effect so that the bed behaves like a liquid, and furnishes a very effective medium for carburizing and carbonitriding.

A variety of processing media are employed, in a range of other furnace types, with controlled atmospheres. In gas carburizing, components are treated in atmospheres produced either by the partial combustion of a fuel-gas/air mixture, giving 'endothermic' gas, or directly within the furnace by the thermal decomposition of synthetic mixtures, for example methanol with nitrogen.

A common variant of carburizing is carbonitriding, carried out at 850–900°C with additions of ammonia to the carburizing atmosphere. This promotes the diffusion of nitrogen as well as carbon into the steel surface, enhancing the degree to which the steel is case-hardened. The nitrogen forms the very hard compound iron nitride and, as will be explained, nitrogen enters into several other heat treatment processes. Nitrocarburizing is exploited to upgrade the performance of low carbon steels, particularly in sliding wear conditions.

LOW TEMPERATURE TREATMENTS

Nitriding is a method of improving the surface hardness of alloy steels that have previously been hardened and tempered. The process temperature is only around 500°C, so components such as large gears and crankshafts are less liable to distortion than when treated in a high temperature process. In gas nitriding, parts for treatment are heated in ammonia, which partially dissociates into hydrogen and nitrogen. The nitrogen diffuses into steels which contain chromium, vanadium, molybdenum or aluminium, to form alloy nitrides with high surface hardness. The process is fairly slow: a nitrided case of about 0.7 mm is achieved after about a hundred hours of treatment.

Plasma nitriding uses the energy of an electrical discharge to bombard the surface of components with nitrogen ions at similar temperatures to those used in conventional nitriding. These treatments involve fairly long cycle times, and are correspondingly more expensive than carburizing case-hardening.

A short-cycle derivative of nitriding, 'ferritic nitrocarburizing', is employed to upgrade the performance of low alloy steels. Carried out at temperatures around 570°C, typically for two hours, the process enriches the surface with carbon and nitrogen, but, in contrast to carbonitriding, the nitrogen predominates. It produces a thin compound layer of iron nitride that enhances wear and corrosion resistance, supported by a nitrogen diffusion zone which, if parts are quenched after treatment, improves fatigue strength. Nitrocarburizing can be conducted in a variety of processing media, with ammonia-rich controlled atmospheres. There are many tradenames, including Tufftriding and Sulfinuz for salt-bath treatments, and Nitrotec, Nitroc and others for gaseous treatments.

SELECTIVE TREATMENTS

The case-hardening treatments described above can be defined as 'thermochemical' processes, involving a combination of thermal treatment and chemical reaction. The third important category, relying on thermal input only, is represented by the induction and flame-hardening processes used in industry to harden selected areas of engineering components such as gears, crankshafts and camshafts. The treatments involve heating the surface of medium carbon or alloy steel components to a temperature at which the carbon enters into solid solution in the iron. This is known as austenitizing, and can be understood by reference to *Fig. 40*: the temperature at which austenite is formed depends on the carbon content. The hot steel is subsequently quenched in oil or water to change the structure near the surface to a martensitic condition (discussed on page 135), thus providing a very hard surface.

High energy beams such as lasers, electron beams and ion beams are now part of the heat treatment repertoire. An ion beam 'fires' high energy atoms at the part under treatment, and can be scanned across its surface. Electron beams are high energy pulses, providing a source of concentrated heat. One of the first applications of electron beam hardening was installed at Chrysler's Kokomo plant, where clutch cams were selectively hardened to a depth of 1.5 mm on eight selected areas at the rate of 250 per hour.

Lasers are not suitable for bulk heating, but they are very efficient at rapidly heating small selected areas. Consequently they are used for hardening, by moving a laser beam across the surface of a steel component. A thin surface layer is rapidly heated to above the temperature at which austenite is formed. Once the beam has moved on, the surface layer cools rapidly by conduction of heat away into the cold bulk of the component.

DEPOSITION OF ULTRA-HARD COMPOUNDS

The steels used for tools and dies have been a focus of new treatments which impart ultra-hard surface compounds such as borides, nitrides and carbides. A well-established example is boronizing, in which boron, derived from boron carbide, is diffused into the surface of the steel for several hours at 900–1000°C, providing a very hard layer of iron boride about 0.15 mm thick.

The Toyota deposition (TD) process, developed in Japan, produces compound layers of carbides of vanadium, niobium and chromium on the steel, which is treated in molten salts at a temperature up to 1050°C. The physical vapour deposition (PVD) process is being exploited to improve the performance of cutting tools by imparting a very thin (about 0.003 mm) coating of gold-coloured titanium nitride, which is extremely hard.

Modern heat treatment technology permits industrial processes to be conducted on a sound scientific basis, often with sophisticated controls incorporating computer systems. These have been developed to integrate process and furnace management control, using digital controllers and microprocessors. Recently the industry has adopted fully computerized automatic mechanical handling equipment. It remains a fertile field for innovation, driven by the need to enhance product quality, conserve energy, contain costs and comply with the ever-increasing demands for reliability and quality.

The science of surface hardening has come a long way since a mixture of viper flesh and mummy dust was used by the ancient Egyptians for carburizing iron.

14
ALLOY STEELS

A thin piece of steel can be quenched and will give uniform hardness throughout its section, but if a massive block is heated and quenched the interior will cool slowly, despite the rapid cooling on the outside. Thus there is a gradual decrease of hardness from the outside to the centre of the quenched block. Furthermore, the unequal dimensional behaviour of the inside and outside causes internal stresses, so that a heavy block of high carbon steel might crack if it were drastically quenched. These difficulties are overcome, and many other advantages conferred, by the use of alloy steels. During this century they have entered more and more into the manufacture of highly stressed components or parts that have to work in a corrosive environment, from armour plate and beer casks, right through the alphabet at least as far as washing machines and yachts. There are many alloy steels with only a few per cent of added elements; careful choice of composition and heat treatment gives useful properties at economic cost, often permitting weight savings to be made. At the other end of the scale, some alloy steels contain 30 per cent or more of added elements.

In order to get the full value of improvements conferred by the alloying elements, many of these steels are used in the heat treated condition. Carbon is an essential constituent of the heat treatable steels – it makes hardening and tempering possible; the alloying elements modify the effects of the treatment.

One of the earliest and most famous alloy steels contained nickel (discussed on page 190). Nickel steels were later improved by the addition of chromium, but they proved to be somewhat temperamental in tempering: they developed a mysterious embrittlement, but a small addition of molybdenum was found to cure this 'temper brittleness'. Another landmark in alloy steel history was the discovery of manganese steel by Robert Hadfield in 1882. Such a steel, containing about 1 per cent carbon and 13 per cent manganese, can be brought to a high degree of toughness by heating it to 1000°C and quenching in water. In this state the structure of the steel is austenitic, and is only moderately hard, but any attempt to cut or abrade the surface results in the local formation of hard martensite, so that the steel cannot be machined by ordinary cutting tools. Manganese steel is therefore used for purposes which require intensely hard metals, such as parts of rock-breaking machinery and railway crossings. It is used for the bars of some prison cells, because manganese steel makes it almost impossible to escape by the traditional method of using a file.

STAINLESS STEELS

In 1912 Harry Brearley, head of the Firth-Brown research laboratories in Sheffield, was asked to direct an investigation on the possibility of improving the erosion resistance of rifle barrels. He experimented with a number of steels with chromium contents ranging from 10 to 20 per cent, checking their microstructures and mechanical properties. In August 1913 he tested a steel with 12.86 per cent chromium, 0.24 per cent carbon and 0.44 per cent manganese. He was surprised that the etching liquids used to prepare specimens for examination did not attack the steel; and, among samples left in the laboratory for twelve days, the chromium steel was still bright but ordinary carbon steel samples were tarnished. Unfortunately the research did not lead to any improvement of rifle barrels, but Brearley wrote, 'These materials would appear specially suited for the manufacture of spindles for gas and water meters, pistons and plungers in pumps, or perhaps for certain items of cutlery.'

During the First World War, stainless steels were used to manufacture inlet and exhaust valves for aircraft engines, followed by a multitude of other applications. The corrosion resistance of stainless steels is due to a naturally occurring chromium-rich oxide film, invisible and submicroscopically thin; if this film is damaged, it is self-healing and protective.

The stainless steels, which now often contain other elements besides chromium, are among the most corrosion resistant alloys available, and their use has trebled in the last decade. Now over 10 million tonnes are made each year; Western Europe makes about 3 million tonnes per annum, Japan over 2 million tonnes and the USA about 1.5 million tonnes. At least two hundred different compositions of stainless steel are known, and about a hundred of them are in regular commercial production. They may be divided into the following groups, classified according to their microstructure: ferritic, martensitic, austenitic and duplex.

1 Non-hardenable stainless steels with 11–17 per cent chromium and less than 0.1 per cent carbon are known as ferritic. They cannot be hardened by heat treatment, but can be strengthened by cold working. Those with 11 per cent chromium are used for vehicle exhausts; those with 17 per cent are suitable for kitchen equipment. The corrosion resistance of these steels can be further improved by adding about 2 per cent molybdenum, making them suitable for marine applications.

2 Hardenable 12 per cent chromium steels, usually containing less than 1 per cent nickel and about 0.15 per cent carbon, are known as martensitic. They can be cold formed in the fully softened condition, and then heat treated to obtain a condition similar to the martensitic structure described in Chapter 13. They are used for knives and industrial cutting tools.

3 Non-hardenable austenitic steels represent about 80 per cent of the stainless steel market. Best known is '18–8', the steel with 18 per cent chromium, 8 per cent nickel and less than 0.1 per cent carbon. These austenitic stainless steels combine excellent corrosion resistance with good formability and weldability. Typical uses include chemical plant and the tankers which transport chemicals.

4 Duplex stainless steels contain 18 to 26 per cent chromium and 4.5 to 6.5 per cent nickel. Their composition is designed to combine the properties of ferritic and austenitic steels. During the 1980s, some duplex steels were further improved by the addition of about 3 per cent molybdenum, giving great resistance to the corrosive conditions in oil refining and sea water piping.

In addition to these four groups, there is a range of other materials, including non-magnetic stainless steels. By adding elements such as copper or aluminium with titanium to steels containing 17 per cent chromium and 4–8 per cent nickel, heat treatment involving precipitation hardening (described on page 159) can be applied, giving great strengths of up to 1500 newtons per mm^2. Such steels are used for aircraft undercarriages and honeycomb stiffened structures for the walls and skins of spacecraft.

After the Second World War, the manufacture of stainless steels began to spread worldwide. Then they were produced by melting steel, partly scrap, in electric arc furnaces, adding the required alloying elements as ferro-alloys and casting the steel into ingots. Now, thanks to the great developments in continuous casting, manufacturers are able to supply stainless steel in a condition much nearer to the finished product. Electric arc furnaces have the disadvantage that the high temperature causes some loss of chromium, but this problem has been overcome by what is known as argon–oxygen decarburization. First the metal is rapidly melted in an electric arc furnace, but not at such a high temperature that chromium loss occurs. Then an argon–oxygen mixture is injected into the molten metal, which refines it and prevents loss of chromium.

The largest consumers of stainless steel are the domestic equipment and automotive industries, but there are many other important applications. Stainless steels have become essential in the construction of oil platforms. They are safer than painted mild steel, since the hazards of toxic fumes and smoke are eliminated, and no corrosion allowance is required. Stainless steel module walling is up to 60 per cent thinner than mild steel walling, which results in significant weight saving and therefore cost saving for the 'float out' operation. Today, more than half the North Sea oil fields employ stainless steel for module wall cladding.

Stainless steel is counter-attacking aluminium in the competition for the manufacture of beer barrels. At present there are about 7 million aluminium and only one million stainless steel barrels in Britain. The aluminium ones are less expensive, but they have to be given an internal coating to prevent corrosion, and this has to be renewed every few years. Stainless steel barrels do not require such maintenance. Another advantage is that the high melting point is a deterrence to thieves, while aluminium is easily remelted for illicit sale. However, the supporters of aluminium contend that steel barrels are noisier and less resistant to impact damage than aluminium, so work is continuing to establish a code of practice for further improvements in the efficiency and performance of the stainless steel competition.

Stainless steels are being used extensively in the building industry. New buildings have included large quantities of stainless steel curtain walling, decorative panels and door and window frames. Their use in the Thames Barrier will be discussed on page 229, and in nuclear engineering on page 257. Stainless steel flue gas desulphurization units are installed by the coal industry in their fight against acid rain from power stations. As science and technology advance, with supersonic aircraft and flights into outer space, further use will be made of stainless steels because they provide such an excellent combination of resistance to heat and corrosion, plus maintenance of strength at high temperatures.

HIGH SPEED STEELS

Before 1900, a cutting speed of 10 metres per minute was considered good for ordinary carbon steel tools; great astonishment was caused at the Paris Exhibition that year when the US Bethlehem Steel Corporation exhibited cutting-tool steels that would cut for hours at 50 metres a minute, and continue to cut even when the speed was increased so much that the tip of the tool became red-hot. The new type of steel, containing 18 per cent tungsten, about 5 per cent chromium and 0.7 per cent carbon, was developed by Fred W. Taylor and Maunsel White. Its remarkable stability at high temperature was largely due to the presence of tungsten carbide and some chromium carbide.

The high speed steels used for cutting tools illustrate the conservation of alloying elements which became necessary during the Second World War and which has become a matter of present concern, owing to the shortages and high prices of many elements used in the manufacture of alloy steels. It was found that high speed steels could be made with less alloying content than was formerly considered essential. Steels containing 3 per cent each of tungsten, molybdenum and vanadium were introduced in Germany; other steels, developed in Britain and the USA, contained 0.8 per cent carbon, 6 per cent tungsten and 4 per cent molybdenum. These alloying contents are to be compared with the 14–22 per cent tungsten which was the normal pre-war range of composition. Raising the vanadium content to as much as 4 to 5 per cent has proved successful in some high speed steels. One of the best metallic cutting materials turned out to be tungsten carbide blended with about 10 per cent cobalt – a material which is all alloying elements, and contains no iron.

NEW DEVELOPMENTS IN ALLOY STEELS

More than half the alloy steels now regularly used in automobiles were not available as recently as 1986. The consumption of alloy steel is steadily increasing, linked with efforts to economize in the use of some rare and expensive metals which are essential constituents in many of the new steels. Fortunately it has been found that serviceable steels can be obtained by adding the alloying elements in only small amounts, linked with new methods of heat treatment.

The high strength, low alloy (HSLA) steels contain about 0.07 per cent vanadium, sometimes with smaller amounts of niobium and titanium, the carbides of these elements being dispersed through the steel, providing a considerable increase in strength. The HSLA steels are used for bridges, high-rise buildings, aircraft hangars, oil and gas pipelines and earth-moving equipment.

Another class is the dual-phase steels; a typical composition is 0.07 per cent carbon, 1.6 per cent manganese and 0.85 per cent silicon. Their structure is of an alloyed ferrite matrix, hardened with between 10 and

20 per cent martensite. These steels are playing an important role in the making of car body panels and chassis members, where high strength and rigidity must be combined with ductility to enable press forming operations to be carried out efficiently.

Stringent control of manufacture and the demand for optimum mechanical properties have led to the development of superalloy steels, completely free from non-metallic inclusions. Such inclusions may originate from refractories with which the steel is in contact in melting and casting; they are potential sources of failure when severe stresses are encountered, as in modern aero engines. The consumable-arc melting technique, used for molybdenum and titanium (illustrated in *Plate 25*), is now being adopted for melting superalloy steels. Such furnaces can melt up to 8000 kg. Vacuum melting is being used increasingly to improve the quality of steels. This technique helps to prevent the formation of non-metallic inclusions, and it eliminates deleterious gases.

A new family of what are called medium carbon microalloyed steels was first developed in Germany, followed by Japan and the rest of the industrial world. The essential feature of these materials is that the amount of added elements is small, but greatly enhanced properties are obtained by tailoring the structure of the steel by suitable processing and heat treatment. Vanadium has been the most popular addition, but niobium has also been used.

Titanium has always been noted as a grain refiner in many alloys; small additions of titanium reduce the grain size and improve the toughness of the microalloyed steels. A typical material, introduced from Japan, contains 0.32 per cent carbon, 1 per cent manganese, 0.12 per cent vanadium and only 0.024 per cent titanium. Since the early 1970s, microalloyed steels have been widely used for crankshafts, connecting rods, gear shift forks and levers. Volvo are using 25 000 tonnes of microalloyed steel a year. The Rover Group in Britain claimed an annual saving of £500,000 by replacing a heat treated manganese–molybdenum alloy steel crankshaft with a microalloyed steel forging.

15

CAST IRON

Hot iron from the blast furnace is generally converted into steel, but a sizeable amount is cast into ingots which, together with scrap steel and scrap cast iron, are remelted in cupolas which look like small blast furnaces and produce cast iron. Coke fuel provides the heat for melting and increases the carbon content of the metal. The coke, iron ingots and scrap are charged near the top of the cupola; air, enriched with oxygen, is blown in around the bottom. As the technology of melting and the rapidity of chemical analysis have improved, it has been possible to increase the proportion of scrap while maintaining the high quality of the cast iron.

Environmental requirements for fume-free production have led to the development of electric melting of cast iron. Scrap iron and steel are melted and carbon is added in the form of carburizers such as graphite or petroleum coke. The two methods have their advantages and limitations, and the choice is still in the balance. Cupolas are more energy efficient than electric furnaces, and the maintenance and refractory costs are lower. Electric melting permits more accurate compositional control than cupolas, and allows close control of pouring temperatures. In Britain about a quarter of the cast iron produced is melted in electric furnaces; in the USA the proportion is about fifty–fifty. In southern Germany cupolas are replacing electric furnaces because of the local high cost of electricity.

The composition of cast iron depends on the requirements of the customer, but a typical material contains about 3.3 per cent carbon, 2.3 per cent silicon and 0.7 per cent manganese. The great fluidity of cast iron and its low shrinkage on solidification make possible close limits of accuracy and considerable freedom in design. Cast iron is easily machinable and has high hardness, compressive strength and resistance to wear. It is the cheapest manufactured metal, with a vast range of end products. Although some small components such as parts of woodworking tools are made, many large castings are produced. So far as we can ascertain, the largest castings are cylinders for paper-making machines, weighing about 50 tonnes.

Cast iron brake discs provide just one example of the research and development carried out by the industry. Most manufacturers of automobile brake discs require the cast iron to provide maximum thermal conductivity. Executive vehicles, where costs are of secondary importance, use a high carbon cast iron to give maximum thermal conductivity, but the brake discs have to be serviced and replaced more often than those on family cars, which have lower conductivity but greater strength and are therefore more durable. Some manufacturers specify alloying additions, ranging from 0.03 per cent titanium to 1 per cent copper, to influence the frictional characteristics, corrosion resistance and noise of the brakes. The automobile brake industry uses substantial amounts of 'permanent mould cast iron' for wheel cylinders, tandem master cylinders and other components. Molten cast iron is poured into permanent cast iron dies whose surfaces are treated with a thin refractory coating supplemented by a layer of soot deposited from an acetylene flame. The castings are then heat treated.

There has been a recent resurgence in the use of cast iron in the streets of towns and cities. While the cast-iron splendours created by our Victorian forefathers may never be repeated, many towns are now installing well-designed castings with an old-fashioned look, as bollards, lamp standards, drinking fountains and bandstands, and a magnificent Chinese gateway in Gerrard Street in London's Soho.

Several varieties of cast iron can be produced by selecting different grades of pig iron and scrap, by varying melting conditions and by using special alloying additions, but in general the two main classifications are white cast irons and grey cast irons. Both contain between 2.5 and 4.0 per cent carbon, but the difference lies in the condition of the major portion of carbon in the structure of the metal. In white cast iron all the carbon is present as cementite, Fe_3C, and the fracture surface of such a material is white. In grey cast iron most of the carbon is present as flakes of graphite, with usually a remainder in the form of pearlite; the fracture surface of this type is grey.

Thomas Turner, one of the great metallurgists of the nineteenth century, who later became the first Professor of Metallurgy at the University of Birmingham, published a historic article entitled 'Influence of silicon on the properties of cast iron' in the *Journal of the Chemical Society* in 1885. His research was a landmark in what we described in the first chapter as the art of metalworking growing into the science of metallurgy. Professor Turner proved that the amount of silicon present determines the condition of cast iron. With only 1 per cent silicon the iron tends to be white, while with about 3 per cent even rapidly cooled cast irons are grey. Later it was found that the presence of other alloying elements also affects the structure of cast iron; for example, chromium tends to produce a white cast iron, and nickel a grey one. Alloying elements are added to some cast irons to improve their resistance to wear or to corrosion.

Since cementite is intensely hard, white cast iron is hard and durable, though brittle. Grey cast iron is softer, readily machinable, less brittle and suitable for sliding surfaces because graphite is soft and is a good lubricant. The rate of cooling to some extent determines whether the iron is white or grey: the more rapid the cooling, the greater the tendency to form a white iron. Use is made of this fact in producing large cast-iron rolls for rolling mills, which have a grey centre but a chilled, and therefore a hard, white iron surface.

White cast iron is used to a much lesser extent than grey, although its hardness and wear resistance make it suitable for components of mining and quarrying machinery. However, it is an important stage in the manufacture of malleable cast iron, which is discussed later.

SPHEROIDAL GRAPHITE CAST IRON

Since 1948, spheroidal graphite cast iron, also known as ductile iron or SG iron, has been developed as a structural material to compete with steel castings or forgings for many stressed components. In the search for improved properties, it was discovered that the addition of about 0.04 per cent magnesium or cerium to molten cast iron caused the graphite to form into small nodules when the metal solidified. Cast iron treated in this way is stronger, tougher and more shock resistant than ordinary cast iron, in which the graphite flakes take up a comparatively large volume. It is necessary that the sulphur content of SG iron is less than 0.02 per cent. A large amount of SG iron is centrifugally cast to make pipes of diameters ranging from 50 to 1600 mm for carrying water and gas at both high and low pressure.

The automobile industry is always on the lookout for cost savings without loss of quality, so SG iron is used for crankshafts and nearly every modern car has an exhaust manifold in this material. Another development is for the callipers of disc brakes, where high quality performance and safety are of paramount importance. Although most parts of the Channel Tunnel have pre-cast concrete linings, SG iron has been used for a large number of underground connecting passages and electrical sub-stations.

Since the late 1970s there has been a further development in SG irons which, although comparatively small at present, seems to be on the verge of major growth. 'Austempered ductile irons' are SG irons alloyed with small amounts of nickel, molybdenum or copper and then heat treated. In the USA several million rail track tiedowns (sleepers), each weighing over 10 kg, have been supplied. The material has also been used for crankshafts of refrigerator compressors. So far automotive uses have been limited, but it is expected that gears, crankshafts and connecting rods will be manufactured in this material.

MALLEABLE CAST IRON

White cast iron is the basis for the production of malleable iron; it is heat treated in such a way that the white cementite is broken down into ferrite and graphite in a structure which tends to be nodular in form and gives a black fracture. Such iron is then known as blackheart malleable iron. *Figure 41* shows the microstructure of the three types of cast iron. In the

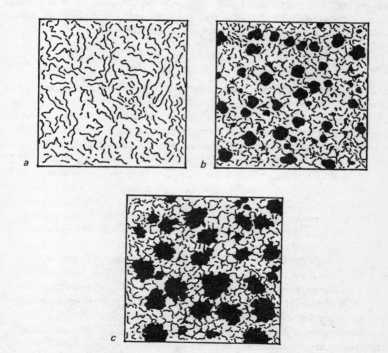

Fig. 41. Microstructure of three forms of cast iron: *a* grey cast iron, *b* spheroidal graphite iron and *c* blackheart malleable iron

grey cast iron the flakes of graphite can be seen. The spheroidal graphite appears as dark, rounded nodules, while the form of the graphite in malleable iron consists of nodules of a fuzzier type than those of SG iron.

Malleable iron production is an old industry, the process having been discovered two hundred years ago. Perhaps it has the benefits as well as the disadvantages of an industry which has been modernized and expanded over the last fifty years. Malleable iron is inexpensive and suited to the needs of modern production requirements because it is strong, tough and can be machined rapidly.

16

ALUMINIUM

The world's annual output of aluminium, known as aluminum in America, is now over 15 million tonnes, to which must be added over 5 million tonnes from recycled scrap. The use of aluminium in aircraft, automobiles, shipping, building and can-making was made possible by bold and imaginative pioneering in the design of components and new methods of fabrication and joining.

Aluminium is a light metal, with a specific gravity of 2.7, only a third that of steel, copper or zinc. It is a good conductor of electricity; for the same cross-section of wire it has about two-thirds the conductivity of copper; but aluminium is better on a weight-for-weight basis because of its lower density. A large number of aluminium alloys can be cast, rolled, extruded, drawn, pressed, machined and welded without difficulty. Many of the alloys can be heat treated to provide high strength. Some are as strong as mild steel, with only one-third the weight.

ALUMINIUM AS A PROTECTOR

The graceful figure of Eros, shown in *Plate 20*, was sculpted by Sir Alfred Gilbert to honour the philanthropist the Earl of Shaftesbury, who died in 1885; it was cast in aluminium by a London foundry. After a hundred years of exposure to the Piccadilly Circus atmosphere, Eros shows little sign of corrosion.* This is because when aluminium is exposed to air a protective film of hydrated aluminium oxide, only a few millionths of a millimetre thick, forms on the surface, sealing the metal and so preventing further oxidation from taking place.

---

* Eros was recently removed for cleaning, removal of graffiti and repair. The crack in one leg and the loose joints were inflicted by human interference; no corrosive damage was discovered. He is now facing in a different direction from that shown in the picture.

The oxide film can be thickened, up to between 0.003 and 0.025 mm, by an electrolytic process known as anodizing. The aluminium articles are suspended in vats similar to those used in electroplating, but generally containing sulphuric acid solution. The components to be anodized are at the positive, anode end of the vat. The action of the electric current releases oxygen from the solution, and a tenacious film of aluminium oxide forms on the surface of the metal. Depending on the operating conditions, anodizing can be adjusted to produce either a film of great hardness or one that is very resistant to corrosion. The anodic film also has the ability to absorb dyes, thus enabling the metal to be tinted with attractive and enduring colours. For special purposes a hard anodic film, up to 0.2 mm thick, can be produced.

ALUMINIUM AS A CONDUCTOR OF ELECTRICITY

An overhead conductor line made of aluminium was installed between Shawinigan and Montreal at the beginning of the twentieth century. Later aluminium conductors were reinforced by a central core of galvanized steel wire, and its use became more general, first in the USA and Switzerland and then in other countries. The resulting strength of the conductors made long spans possible without too great a sag. In Britain, two conductor types have been used in overhead transmission lines: 'LYNX' comprises 30 strands of aluminium in two layers, over a seven-strand steel core; 'ZEBRA' has 54 strands in three layers over a similar steel core.

In the UK, aluminium alloy conductors were used from the 1930s, but their use was limited to the transmission of voltages below 132 kVA. By the early 1970s, France had introduced aluminium alloy conductors for high voltages, and the UK followed suit in 1978 for 275 and 400 kVA conductors. The all-aluminium alloy conductor (AAAC) contains about 0.5 per cent magnesium and 0.5 per cent silicon, and is heat treatable. It has about twice the strength of pure aluminium, with a reduction of only 13 per cent in electrical conductivity.

The use of aluminium alloy on its own overcomes a 'headache' with steel reinforced conductors. Galvanic corrosion, which will be discussed on page 227, can affect the durability of two metals in contact with each other when they are subjected to the ravages of rain, ice and mist. Although the problem has been partly reduced by covering the internal layers of the conductor with grease, aluminium alloy conductors are preferable since galvanic corrosion is absent. A further advantage is that, because they weigh less than steel reinforced conductors, improvements in the power system are achieved. For these reasons the British National Grid Company now employs the 'RUBUS' conductor, consisting of 61 wires, each of AAAC, 3.5 mm thick.

In most parts of the world steel-reinforced cables are used for extremely long spans, and continual efforts are being made to increase efficiency and reduce cost. The original cable across the confluence of the Rivers Severn and Wye consisted of a core of 91 steel strands surrounded by two layers of aluminium alloy, containing 36 and 42 round strands respectively. These conductors suffered from flashing between the phases, and were replaced by conductors with trapezoidal-shaped outer strands, which solved the flashing problem. Another long river crossing is over the River Usk, near Newport in South Wales. This 1.7 km long line was refitted in 1991 with a steel-reinforced aluminium alloy conductor. It consists of a galvanized steel core of 61 strands, surrounded by 90 wires of aluminium alloy, wound in three layers.

SOME USES OF PURE ALUMINIUM

Aluminium foil, sometimes called 'silver paper', starts its life as a rectangular ingot weighing about 12 tonnes. It is heated to between 500 and 550°C and passed through a series of hot rolling mills until it is reduced in thickness to about 3 mm. Cold rolling then reduces it to about 0.45 mm. It is annealed and given further cold rolling, bringing it to 0.035 mm. When further reductions are required, a double rolling technique is applied which brings to the metal a finish that is bright on one side and matt on the other. Typical thicknesses of aluminium foil are 0.008 mm for wrapping chocolates and cigarettes, 0.015 mm for foil for use in the kitchen, 0.04 mm for milk bottle tops and 0.1 mm for food containers.

Convenience foods are mass-packaged with the aid of foil: the mix is put into individual dishes which travel on a conveyor in the cooking oven. Meat and poultry for roasting can be wrapped like a parcel in foil, after being covered with fat; for the consumer no basting is necessary and the foil wrapping retains the juices and reduces the escape of cooking odours.

ALUMINIUM AS A REFLECTOR

Aluminium is a good reflector of heat; storage containers for petrol, milk and other liquids are coated with crinkled aluminium foil which reflects away the sun's heat instead of absorbing it, and so the liquid inside is prevented from becoming unduly heated. Also, for the same reason, aluminium is used in buildings to maintain constant temperatures in hot and cold weather; the efficiency of aluminium foil for insulation purposes depends upon its being associated with a still air space.

Polished or electro-brightened high purity aluminium is one of the best materials for reflecting light, better than ordinary mirror glass, and in some respects more suitable than silver. Aluminium-coated reflectors have two advantages of special interest to astronomers; first, when the front of an astronomical telescope mirror is coated with aluminium rather than silver, it does not tarnish so rapidly; second, aluminium reflects ultraviolet light better than silver does.

The use of an aluminium mirror coat is now standard practice for all large reflecting telescopes, including the 200-inch Hale Telescope on Mount Palomar. This famous telescope is situated in a dry climatic zone, in California; it is re-aluminized about once every five years and is carefully washed every few months. Mirrors in wetter climates deteriorate much more rapidly. For example, when the Isaac Newton Telescope was located at Herstmonceux, near the coast of Sussex, the 2.5-metre mirror lost reflectivity within a few months, due to salt in the air. Therefore it had to be re-aluminized every spring, summer and early winter: the aluminium was stripped off, and the mirror cleaned, polished and dried, with a final holding in vacuum before the fresh coating of aluminium was applied. During the early 1980s the telescope was transferred to the observatory site on La Palma in the Canaries, high on a mountain where conditions are more suitable. Aluminized mirrors can be protected by evaporating a very thin layer of transparent fused silica onto their surfaces. However, such coatings transmit only a limited range of colours. New silica-based materials are now being developed which do not suffer this disadvantage and they are being used for mirrors up to 500 mm in diameter.

ALUMINIUM CASTING ALLOYS

A wide range of alloys are available, with a variety of properties for a variety of uses. Silicon, copper and magnesium are the main alloying elements, either alone or in combination. Often small amounts of other elements are added, including manganese, zinc, titanium and nickel. All these alloys can be cast in sand moulds, and nearly all can be permanent mould cast; a limited number are suitable for pressure diecasting.

The alloy containing about 13 per cent silicon shows a metallurgical phenomenon known as modification. As normally cast in a sand mould, it has a coarse structure (shown in *Plate 31*c) accompanied by low strength and weakness under shock. In 1920, Aladar Pacz discovered that a small addition of the compound sodium–potassium fluoride to the molten alloy brought about a change of structure when the metal was cast. Later it was found that other compounds of sodium, or about 0.05 per cent of the metal itself, added to molten aluminium–silicon alloy would modify the structure obtained on solidification (shown in *Plate 31d*). The alloy treated under these conditions is much stronger and tougher than the unmodified alloy; it has an excellent combination of strength, ductility, resistance to corrosion and castability. A foundry foreman we once knew was heard to remark, 'It runs like milk.'

An alloy known as 380 in the USA and LM 24 in the UK contains between 7.5 and 9.5 per cent silicon and from 3 to 4 per cent copper. It can be produced from scrap and is therefore comparatively inexpensive. It is diecast in vast tonnages, and is more suitable for rapid machining than the 'straight' aluminium–silicon alloy; this property is important in the mass-production industries, where fast machining and long life of cutting tools are essential.

Various specifications are employed in different parts of the world. The alloy with 10–13 per cent silicon, known as LM 6 in the UK, is referred to as A 413.2 in the USA, DIV in Japan, and 4261 in Denmark and Sweden. In recent years international nomenclatures have been agreed upon that indicate the composition of the alloys. Under the ISO specification DIS 3522, the aluminium–silicon alloy mentioned above is designated Al–Si 12. The diecast alloy with about 8 per cent silicon and 3 per cent copper has the specification Al Si8 Cu3 Fe (the Fe indicating that in addition to the main alloying elements a small amount of iron is permitted). This nomenclature is logical but it may take a little time before foundry supervisors in conversation refer to such a 'mouthful' instead of the British LM 24 or the very descriptive American 380.

WROUGHT ALUMINIUM ALLOYS

As wrought products include sheet, foil, bar, wire, forgings, extrusions and tubes, they cannot be classified as simply as the casting alloys. Code letters and figures indicate the form of the product and the type of heat treatment it has been given. The wrought alloy compositions are classified in an internationally agreed system, according to the main alloying

Table 12. Classification of wrought aluminium alloys

Series	Main alloying elements
1000	(Aluminium of over 99 per cent purity)
2000	Copper
3000	Manganese
4000	Silicon
5000	Magnesium
6000	Magnesium and silicon
7000	Zinc
8000	Lithium and miscellaneous other added elements

elements; each has four digits (*Table 12*). There is a 9000 group 'in waiting' for the possible development of new alloys.

In the 2000 range, the alloys in the heat treated condition have mechanical properties similar to those of mild steel – good stiffness and fatigue resistance – but to improve their resistance to corrosion they are often clad with pure aluminium, as will be described on page 160. A well-known alloy in this group, 2014A, contains from 4 to 5 per cent copper, with 0.4 to 1.2 per cent manganese and small amounts of other elements. It is a major aircraft construction material for such parts as wing skins and structural components.

A limited amount of manganese, up to about 1.5 per cent, can be alloyed with aluminium, so there are only a few compositions in the 3000 group. They have moderate strength, but excellent workability and good corrosion resistance. The alloys in the 4000 group, with silicon, have only a few applications in the wrought condition but they provide the backbone of the aluminium casting industry. The 5000 group contains magnesium as the major element but some alloys also include small amounts of manganese. They have good strength, good weldability and corrosion resistance, and are specially suitable for marine applications. The 6000 group, containing magnesium and silicon, can be heat treated and are used for structural parts such as windows and doors which are extruded, anodized and often tinted with colours.

The 7000 group contains up to about 7 per cent zinc with about 2 per cent copper and magnesium, plus small amounts of manganese, chromium and zirconium, and are among the strongest of light alloys. When heat treated they have a strength of up to $600 \, N/mm^2$ and are widely used in the aircraft industry. The 8000 alloys, containing lithium, will be discussed on page 270. The compositions of some alloys used in the aircraft industry are given in *Table 14* on page 160.

AGE HARDENING

Early in the twentieth century, Alfred Wilm, a German research metallurgist, was investigating the effect of additions of small quantities of copper and other metals to aluminium in the hope of improving the strength of cartridge cases. He tried various combinations and different forms of heat treatment until, more or less by accident, an extraordinary discovery was made.

An aluminium alloy containing 3.5 per cent copper and 0.5 per cent magnesium was heated, then quenched in water and tested. The results were not particularly impressive. A few days later, some doubt was expressed about the accuracy of the tests, so pieces from the same batch were tried again. To Wilm's surprise, the hardness and strength were much higher than they were before; this led him to perform a series of experiments on the effect of storing the alloy for different periods after heating and quenching. The strength gradually increased to a maximum in four or five days, and the phenomenon became known as 'age hardening'. In 1909 Wilm gave to the Dürener Metallwerke at Düren sole rights to work his patents; hence the name 'Duralumin'.*

The Zeppelin engineers realized that the design of airships might be revolutionized by the use of this light, age hardened alloy. The rolling of the alloy into sheets and strips presented a number of problems, but intense efforts were made so that most of the obstacles were overcome, and during the First World War large amounts of age hardened aluminium alloy, strip, sheet, girders, rivets and other parts were used, first for Zeppelins and afterwards for other types of aircraft.

Since then many age hardening aluminium alloys have been developed. It was found that the strength of an aluminium alloy containing copper, nickel and magnesium could be increased by 50 per cent. This provided the basis of several alloys developed by Rolls-Royce in association with aluminium producers. A typical alloy (specified as 2018A), containing 2.5 per cent copper, 1.5 per cent magnesium, 1.2 per cent iron and 1.0 per cent nickel, was used for the fuselage covers of Concorde.

---

* Although 'Duralumin' was the name given to the age hardening alloys developed by Wilm, the title is now a registered tradename for a range of alloys belonging to British Alcan Aluminium Ltd.

THE MECHANISM OF AGE HARDENING

Metallurgists all over the world attempted to explain the mechanism of age hardening, but it was not until some years after its discovery that the problem was even partially solved, although hardening by quenching and ageing was used extensively in the meantime. Metallurgical detective work in revealing the causes of age hardening was slow because the instruments then available, including the metallurgical microscope, were not sufficiently sensitive to follow the minute changes in the internal structure of the alloy. Indeed, for many years it was not certain which of the impurities or added metals was responsible for age hardening. This was because, even when viewed under the microscope, there was no visible difference in the structure of the treated alloy before and after ageing. Some changes in structure could be detected by X-rays, but it was not until the advent of the electron microscope that it became possible to explain the mechanism of age hardening.

When the solid alloy with 4 per cent of copper is at a temperature of 500°C, the copper enters into solid solution in the aluminium (see page 49). At room temperature, however, aluminium can hold less than 0.5 per cent copper in solid solution. When this alloy is quenched in water from 500°C down to room temperature, the copper does not immediately come out of solid solution.

Over the period of some days after the alloy has been quenched, the copper atoms, which can no longer be held in solid solution, are forced to move or diffuse among the aluminium atoms to form minute areas containing higher amounts of copper than the average. These areas ultimately form the intermetallic compound $CuAl_2$ (page 51). This delayed effect may be compared with the behaviour of some kinds of home-made jam in which sugar crystallizes out, apparently of its own accord, when the jam is left to stand for a long time. The reason is that when the jam is hot it will hold more sugar in solution than when cold; the sugar does not crystallize immediately the jam is cold, but becomes apparent several months afterwards.

All this describes *what* takes place inside the alloy. *Why* the local aggregations of $CuAl_2$ cause age hardening to occur was controversial for more than fifty years after Wilm's discovery. Gradually it has been realized that the hardness and strength of metals are related to resistance to slip of the crystals and increases in lattice strain. All the research on 'dislocations' described on page 60 has added to the evidence that the hardening is produced by the formation of areas or zones of copper and aluminium atoms, which hinder further slip and distortion.

Table 13. Effect of different treatments on hardness of an aluminium–copper–magnesium–silicon alloy

Treatment	Brinell hardness (approximate)
1. As quenched	60
2. Quenched and kept at room temperature (i.e. naturally aged)	120
3. Quenched and reheated to 175°C (i.e. precipitation treated)	150

PRECIPITATION TREATMENT

We have seen how, in an aluminium–copper alloy, the intermetallic compound $CuAl_2$ is precipitated at room temperature and thus increases the hardness of the alloy. The hardening can be hastened and intensified by heating the quenched alloy at about 175°C. This is known as 'precipitation treatment'. Furthermore, if magnesium and silicon are also present a compound of these two elements, Mg_2Si, is formed and takes part in a similar process of hardening to that occurring in ageing. When the aluminium–copper–magnesium–silicon alloy is quenched and reheated to 175°C, particles of Mg_2Si are precipitated in addition to $CuAl_2$. The alloy then becomes harder than if it had been aged at room temperature, as is shown in *Table 13*.

This type of hardening has since been discovered in alloys of other metals, such as magnesium, copper, zinc, tin, lead and iron. Many of them will not harden by ageing at room temperature. In the process, the alloy is first raised to a fairly high temperature at which a proportion of the appropriate alloying element goes into solid solution; this is known as 'solution heat treatment'. After a rapid quench, the alloy is precipitation treated to make the dissolved element come out of solution in minute agglomerations of such a size that maximum hardness is reached.

Alloys which age at room temperature are known as age hardening alloys, while those requiring treatment at higher temperatures are called precipitation hardening alloys. The essential difference between the precipitation treatment of such alloys and the heat treatment of steel is that steels attain their maximum hardness on quenching, while tempering usually *reduces* hardness. Precipitation treated alloys, as quenched, are comparatively soft, but the treatment *increases* the hardness.

Table 14. Composition (per cent) of some aluminium alloys used in the Airbus

Element	Alloy				
	2024	2618	7010	7075	7150
Copper	3.8–4.9	1.8–2.7	1.5–2.0	2.0–2.6	1.9–2.5
Manganese	0.3–0.9	0.25	—	0.1	0.1
Magnesium	1.2–1.8	1.2–1.8	2.1–2.6	1.9–2.6	2.0–2.7
Zinc	0.25	0.15	5.7–6.7	5.7–6.7	5.9–6.9
Zirconium	—	0.20	0.15	—	0.08–0.15
Titanium	—	—	—	0.06	0.06

ALUMINIUM IN AIRCRAFT

There have been aircraft in which most of the fuselage was of wood – the Mosquito, for example – but as early as 1903 the famous aeroplane of the Wright brothers contained several parts made of aluminium. Although advanced military aircraft now contain a large amount of titanium alloys, most commercial aircraft rely on aluminium alloys for over 70 per cent of their weight. The choice of alloy depends on the stresses to be encountered, the temperature at maximum speed and the corrosion resistance requirements.

Table 14 lists some of the alloys in the European Airbus. The permitted ranges of the principal elements are given; where small percentages of other metals are included, they are given as a single figure, indicating the maximum. Alloy 7075 is used for the Airbus frame and tail, 2024 for the surface skin, 7010 for the wing spars and wing box, and 7150 for the top and 2024 for the bottom wing surfaces.

ALUMINIUM CLADDING

A large proportion of the weight of an aircraft consists of heat treated aluminium alloy in the form of thin sheets which cover the fuselage and wings. Some of the aluminium–copper alloys in the 2000 range, though very strong, do not have good resistance to corrosion. To provide the necessary protection, and to retain the strength, a process known as cladding is used. A thin layer of pure aluminium is rolled on to each side of the alloy sheet, making a three-ply metal. This combination of protective outer coating and strong alloy centre can be exposed for over five years to the continuous action of a salt spray test without showing any deterioration through corrosion. The thickness of the aluminium coating varies according to the gauge, but as a general rule the total thickness of the aluminium, front and back, amounts to 10 per cent of the thickness of the sheet. The rolled clad material is heated to about 500°C, then quenched and thus brought to the soft condition. It is then pressed or beaten into the required shape. After that the clad metal is aged in a temperature controlled furnace, where the alloy core of the clad sheet increases in hardness and strength by precipitation hardening.

Many of the Boeing aircraft operated by American Airlines make a feature of their silvery appearance. This provides scope for clad alloys to be used. Other airlines, including many in Europe, colour their aircraft, so the sheets of aluminium alloy are anodized and then painted; cladding is not necessary.

HONEYCOMB TECHNOLOGY

Honeycomb panels have become a feature of aircraft body construction, and have spread to many other areas – building, boats, racing cars, skis and other sports products. Alloy foil between 0.04 and 0.08 mm thick is converted into hexagonal cells about 5 cm across. A typical construction is shown in *Fig. 42*. They are generally made from the aluminium–manganese alloy 3003; outer skins, of either aluminium sheet or plastic, are secured to the honeycomb by adhesive. These honeycomb panels have excellent compressive and shear strength, and combine light weight with stiffness.

ALUMINIUM IN SHIPS

As long ago as 1890, Alfred Nobel saw the possibilities of the new metal aluminium, and used it to make *Mignon*, his 12 metre boat, which could carry 25 passengers and cruised on Lake Zurich. The advantages of aluminium in improving stability and making possible larger super-structures came to fruition soon after 1945. Aluminium was first used in small vessels, including lifeboats; among later landmarks in the development of aluminium for shipping were the SS *United States* in 1952, and then the *Oriana*, *Canberra* and *France*, each of which contained about a thousand tonnes of welded aluminium alloy in their structures. Other industries contributed to the development of aluminium in ships; for example, welding the metal became much more practicable when the argon arc process was introduced (page 239).

Confidence in the use of aluminium led to the choice of aluminium–magnesium alloy for the four decks above the steel portion of the

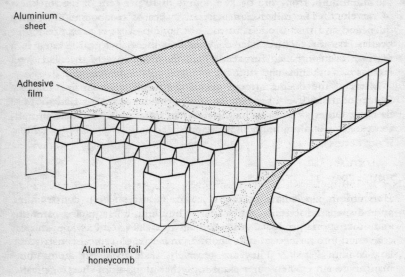

Fig. 42. Aluminium honeycomb sandwich panel

Queen Elizabeth 2. The additional cost of the alloy was about £500,000, as compared with the basic cost of the ship of £25 million, so the increased cost of using aluminium was about 2 per cent, amply justified by the increased revenue potential from the expanded carrying capacity.

ALUMINIUM IN BUILDINGS

Many buildings constructed since 1945 have employed aluminium curtain walls, the metal being either commercially pure or with only a small amount of alloying elements. The curtain walling of skyscrapers may require about 1000 tonnes per building. In London, the headquarters of the *Daily Telegraph* has a design resembling a ship, in keeping with the nearby dockland environment. The upper parts are of aluminium with about 0.5 per cent of each of magnesium and manganese. Profiled lengths 7.5 metres long were produced from coils which had been etched in caustic soda, treated with chromate and then coated with a polyvinylidene fluoride paint.

ALUMINIUM IN OIL PLATFORMS

The development of aluminium technology for offshore structures was led by Norway, with what is termed 220-man five-storey accommodation modules. These structures, made of aluminium alloy, weigh 2000 tonnes, including beams, walls, floors and roofing. Oil rig topside structures weigh over 20 000 tonnes, and the need to reduce weight has become increasingly important as more and more equipment is installed to enhance oil recovery. This has led to the use of aluminium alloys for helicopter decks, walkways, gantries, bridges and many other essential features.

ALUMINIUM CANS

Nearly 150 thousand million aluminium cans are made each year, and the number is growing as more and more countries adopt this popular way of dispensing soft drinks and beer. The aluminium can is a splendid way of saving energy. It can be recycled, which takes only five per cent of the energy required to smelt the metal from bauxite. Every year sees technical improvements to increase productivity, economize material and make the design of cans even more acceptable to users. The savings in transport are considerable. A bottle weighs 230 g, and tinplate containers of similar capacity weigh about 35 g, but an all-aluminium can weighs only 20 g. Worldwide, four out of every five beverage cans are made of aluminium, though it must be added that the ever-enterprising steel and tinplate industries are fighting back.

The canmakers receive coils of rolled aluminium alloy weighing about 10 tonnes. Then, by a long series of shaping operations, and with fabulous productivity, the bodies which will hold the drinks are made. In the USA, sheet 0.3 mm thick is formed into cups, 14 at a time, with a single stroke of the press. This is followed by a process which forms the finished shape at 300 strokes per minute. The easy-open tops are blanked from strips of aluminium alloyed with 4.5 per cent magnesium. It is rolled into strips about a third of a millimetre thick and blanked. Each top is V-grooved to form the tab which will be pulled to open the can.

The stay-on tab is growing in popularity, featuring an integral rivet similar to that on the ring-pull can; the tab is raised to push in the pouring aperture but remains attached to the can, thus helping conservation of material.

ALUMINIUM IN AUTOMOBILES

By extensive use of aluminium alloys the power-to-weight ratio of many car engines has been massively improved. The great attractions of aluminium compared with cast iron are light weight and good heat conductivity; the latter means that less engine coolant needs to be used, which in turn provides a rapid warm-up for starting.

A pioneering development in the USA during the 1950s led to the aluminium alloy cylinder block, which proved that other large and complex components could be made in aluminium. The first diecast cylinder blocks were produced in Toledo. Cast iron cylinder liners were placed in the die, and the aluminium alloy was cast around them. Although this early project was discontinued after a few years, much experience in design and production technology had been gained for future progress. In the early 1990s, General Motors introduced their 4.9 litre V8 Northstar engine, featuring a diecast cylinder block with a 'cast-in' cylinder liner. It is produced in the 380 alloy, with 8 per cent silicon and 3 per cent copper.

Over thirty years separates the first cylinder block with the cast-in liner and the production of the modern Northstar engine. During this time, very many aluminium alloy cylinder blocks have been diecast in France, Italy, Japan and other countries, but in all of them the liners were inserted after the block was made; the liners can be extracted and either reworked or replaced when necessary. This avoids the production problems which tend to slow down the process when inserts are placed in the die before casting.

A cylinder block for the Chevrolet Vega was diecast in the 1970s, using what is known as a 'hypereutectic' aluminium–silicon alloy, containing 17 per cent silicon and 4 per cent copper. The silicon content is greater than the 13 per cent of the eutectic – hence the 'hyper'. After a chemical etching treatment this alloy is so abrasion resistant that it is not necessary to insert cast iron liners. The diecast Vega block was produced by General Motors in the late 1970s, but was then discontinued. However, the hypereutectic alloy is now widely used in many industries and worldwide annual production amounts to over 100 000 tonnes. In America its greatest use is for pump bodies and compressor components, and in a variety of small engines for such mass-produced machines as chain-saws. In Europe, engine blocks for Mercedes, Porsche, Audi and BMW cars have been made with the hypereutectic alloy.

Other impressive achievements with aluminium include its use for the complex components of automatic transmission systems. Aluminium alloy wheels began as status symbols on sporty cars, but, with the growing efficiency of low pressure diecasting in Europe and America and the 'pore-free' diecasting process in Japan (discussed on page 76), aluminium alloy wheels are now produced by the million.

In the late 1970s, in their search for wider markets, the aluminium producers were hoping – perhaps praying – for another development like the use of aluminium in ships and buildings. Their prayers were answered when it was decreed that US cars must be made lighter with the object of decreasing fuel consumption by at least a third. In addition to reduction in the size of their cars, US manufacturers began a long-term programme that required more aluminium components to replace cast iron. For example, Ford replaced cast iron intake manifolds with aluminium alloy ones, weighing only 7 kg compared with the 22 kg cast iron equivalent. Other uses of aluminium include brake master cylinders, brake drums, bumpers and cylinder heads. Such developments are expected to increase the amount of aluminium in American cars from about 50 kg in the 1980s to twice that amount in the 1990s. European cars may not show such a great increase because they are lighter than their American big brothers, but in Europe and Japan there is the same trend to weight reduction by using more aluminium. Of course, competition continues. Where lightness is required magnesium is a strong competitor, especially in Germany and the USA. Great progress has been made in the production of cast iron components, and certainly that material has the advantage of low cost.

At present there is hard-fought competition between copper and aluminium for automobile radiators. Before aluminium could be used with confidence, a great deal of research had to be carried out in joining processes and alloy selection. The automotive and aluminium industries concentrated on two methods of assembly: mechanical and brazing. The former is simpler and cheaper, but is generally suitable only for cars with engine capacities of 1600 cc or less. Larger cars require a brazed assembly, which occupies less space than a mechanically joined one. Elaborate equipment is needed for the brazing process, using vacuum and specialized brazing methods such as the Nokolok process, developed by Alcan. The alloys most commonly used are those with manganese, though alloys containing magnesium and silicon have also been specified. In the early 1990s, aluminium accounted for over 80 per cent of the European car radiator market, but recently some manufacturers have moved back to copper and its alloys.

Alcoa in the USA and Audi in Germany have been responsible for an exciting new development – the aluminium space frame. Instead of three hundred or more stamped steel components spot welded together to form the body structure, less than a hundred aluminium alloy extrusions are interconnected by vacuum diecast aluminium alloy nodes, robot-welded to the extrusions to form the body structure (*Plate 21*). Tooling costs are halved, and a weight saving of over 30 per cent is achieved, compared with the traditional steel body. The space frame may change the manner in which other cars are assembled in the future.

17

COPPER

Six thousand years ago, copper was discovered in the Middle East after a wood fire had been lit over a mineral containing copper; later this primitive process was carried out deliberately. Tin was made in the same way in this region. By 3000 BC, early metallurgists were making bronze, the alloy of copper and tin, by melting pieces of the two metals together in earthenware crucibles. They found that different bronze compositions gave different properties. An ancient Chinese book on metallurgy, entitled *K'ao kung chi*, recommended four parts copper to one part tin for axes, and two parts copper to one part tin for knives.

The sixth wonder of the world, the Colossus of Rhodes, was a statue representing Helios, the Sun-god. Maybe it did not bestride the harbour, but it was 70 cubits (about 30 metres) tall. The Colossus was made of bronze, reinforced with iron and weighted with stones. An earthquake destroyed it in 225 BC and it lay in the sea for nearly a thousand years, when Arabs who raided Rhodes salvaged the bronze remains. This amounted to nine hundred camel-loads, and was sold profitably as scrap metal.

Later artefacts displayed the splendid colour and castability of bronze. The twelfth-century doors of the cathedral at Pisa were so renowned that the Calimala, rich Florentine businessmen, decided to finance three pairs of bronze doors for their Baptistery. In 1329 they sent a delegate to study the Pisan bronzes and find the best craftsmen to do similar work. Ten years afterwards, Andrea Pisano's doors were installed in the south portal. Nearly a century later the Calimala announced a competition for the pair of north doors. The 24-year-old Lorenzo Ghiberti won, and his doors were installed in 1424. Without any further competitions, the contract for the east doors was awarded to Ghiberti. These marvellous bronze doors, which Michelangelo said were worthy to be the entrance to Paradise, contained ten bas-relief panels illustrating Old Testament scenes, from Adam and Eve, via David and Goliath, to Solomon and the

Queen of Sheba. They have shown signs of corrosion from the twentieth-century Florentine atmosphere, so were removed for restoration; excellent gilded bronze copies have taken the place of the originals. A few restored panels can be seen in the cathedral museum, where eventually all ten Ghiberti panels will be displayed.

An early use of unalloyed copper was for the anti-fouling sheathing of ships' bottoms. Freedom from fouling played a significant part in Nelson's victories. Later, scientific discoveries and the coming of the Industrial Revolution led to a massive increase in the production of copper. The following chronology, beginning 26 years after the Battle of Trafalgar, shows some highlights in the use of copper in electrical engineering, linked with the career of an Irishman who saw the possibilities of the metal and who was partly responsible for the mining of copper ores to meet the demand.

1831 Michael Faraday discovered the principle of electromagnetic induction, which made possible the development of electric motors and dynamos.

1856 A young Irishman, Marcus Daly, left his home in Ballyjamesduff to seek his fortune in America.

1861 Antonio Pacinotti invented the ring winding system, using copper, which made the electric dynamo a practical proposition. His work was published in an obscure Italian journal, but received little attention.

1862 Daly got a job as foreman of a silver mine in Nevada, and became well known as a mining engineer and manager with an uncanny ability to assess the potential values of metal ore deposits.

1866 A telegraph cable was successfully laid across the Atlantic.

1870 Zenobe Gramme rediscovered Pacinotti's invention, and the dynamo began to be developed.

1875 Michael Healey, prospecting for silver, staked a claim on a hill in Montana. He remembered a newspaper editorial which had said, 'General Grant will encircle Lee's forces and crush them like a giant anaconda.' Healey gave the memorable name to what was later to be called the richest hill on Earth.

1876 Alexander Graham Bell transmitted speech via a copper telephone wire.

1878 Thomas Alva Edison produced his incandescent electric lamp.

1879 The workings of electric dynamos and motors were demonstrated at the Berlin Exhibition, where a small electric locomotive pulled three cars containing twenty passengers.

1880 Michael Healey had by now staked several silver claims near Anaconda, but was needing more capital to expand them.

1881 Marcus Daly, now a prosperous mine manager, met Healey, who offered him a share of the Anaconda property. Daly at once began to deepen and develop these silver mines, but often struck irritating outcrops of copper ore, which nobody particularly wanted.

1882 Edison opened the Pearl Street generating station in New York, to supply electricity for five thousand lights.

1882 Daly discovered rich copper ore at Anaconda and had a hunch that this metal was worth developing.

1882–84 Daly shipped 37 000 tons of rich copper ore to Swansea for smelting.

1883 Daly began to operate a copper smelter in Anaconda. This was in full swing by 1884.

Thus a man of energy and vision was in the right place at the right time, and within a few years the rapid growth of the use of electricity caused a tremendous surge in the production of copper. Apart from the discoveries in the USA, copper ores were mined in other parts of the world and smelted in South Wales, but eventually the more economical way of smelting near the mine was developed and Britain lost her position as centre of the copper industry. Chile and the USA are now, almost equally, the largest producers of copper, followed by Russia and other countries of the former USSR, then Canada and Zambia. Recently mines have been opened and smelting plants built in China, Australia and Portugal.

Copper has a higher conductivity of heat and electricity than any metal except silver. The pure metal is ductile and can be rolled into foil only 0.02 mm thick, drawn into wires less than 0.02 mm in diameter, pressed, forged, beaten or spun into complex shapes without cracking. Such ductility is also possessed by several copper alloys, notably brass.

Copper and its alloys have attractive appearances and colours, ranging from red in the pure metal to ochre, gold or yellow in the various alloys. They can be cast with ease with beautiful results, as exemplified by the bronze castings described at the beginning of this chapter. Copper, and most of its alloys, can be joined by soldering, brazing or welding. It is resistant to many forms of corrosion and when, with the passage of time, copper roofs become tarnished, they develop an attractive green patina, known as verdigris. About 8 million tonnes of copper are produced each year, third in tonnage only to iron and aluminium. About 3 million tonnes of this is used in the electrical industries, and about 2 million tonnes for building and water supplies.

COPPER AS A CONDUCTOR OF ELECTRICITY

In 1837, Charles Wheatstone and William Cooke patented the first electric telegraph, using copper wire. They improved their signalling apparatus, and in 1845 a telegraph line was installed alongside the railway between Paddington and Slough. A message sent over this line helped to catch an escaped murderer. It was not until 1866 that, after several attempts, a cable was laid under the Atlantic. Nowadays about a third of all the copper produced is used in the pure form for carrying electric current, ranging from massive underground cables to thin wire for domestic equipment. It is also used for busbars, switchgear and components of transformers, dynamos and motors.

Long-span overhead cables require high conductivity, but the metal must also be strong to support its own considerable weight and to withstand additional stresses imposed by wind and the accumulation of ice. To give it greater strength, copper can be alloyed with other metals but most of them cause a considerable reduction in electrical conductivity. One element, cadmium, allows an acceptable compromise. When about 0.8 per cent cadmium is alloyed with copper, the strength of the metal is greatly increased but the conductivity is reduced to only nine-tenths that of pure copper. There are still many overhead lines in cadmium–copper, but in recent years 'all-aluminium alloy conductors' (discussed on page 152) have been used. Where maximum conductivity is paramount, as in the Channel Tunnel, pure copper is used for current-carrying cables. There has been competition from other directions: advanced laser and amplifier systems have made it possible for optical fibres to transmit speech, thus replacing copper wires. When someone speaks into a telephone their voice is converted first into electrical pulses and then into a sequence of light pulses by a laser beam. These are transmitted along the fibre to the receiving station, where they are converted back to sound.

COPPER FOR WATER AND GAS

Copper was used by the ancient Egyptians for water conduits as long ago as 2750 BC. Today, the metal's use for carrying water and gas is second only to its electrical applications. In the UK and USA, copper accounts for over 90 per cent of all tubing for carrying drinking water in domestic buildings. This is due to its high resistance to corrosion, coupled with malleability and ease of joining. Copper's resistance to biofouling extends to a high resistance to bacteria in drinking water. This is specially true for the infamous bacterium *Legionella pneumophila*, the causative agent of legionnaires' disease.

BRASS

Brasses containing less than 36 per cent zinc are ductile when cold, and can be worked into complex shapes without the necessity for frequent annealing. A cartridge case for a 7.62 mm bullet is an example of the use of a brass containing 30 per cent zinc. A strip about 3.2 mm thick is cut into circular blanks about 30 mm in diameter, and each blank is pressed into the shape of a shallow cup. By a series of further pressing operations the cup is pushed and squeezed through successively smaller and smaller holes in a steel die, elongating the walls of the cup until eventually a tube with a relatively thick base and thin walls is obtained. Further operations are carried out to the base to form the recess for the detonating cap and ensure that the cartridge fits only one type of breech. The rim of the cartridge is also softened by annealing so that it may be bent in to clip the bullet.

The brasses which contain above 36 per cent zinc are harder and stronger than those containing less. This is noted on pages 58–60, where it is shown that with up to 36 per cent zinc the alloy consists of an alpha solid solution. Between 36 and 42 per cent another solid solution, beta, is also present, and these alloys are called alpha/beta brasses; 60/40 brass is the best-known example. Although such brasses are less workable at room temperature, their plasticity is increased at high temperatures. They are usually shaped by hot rolling, extruding, hot stamping, casting or diecasting.

While the alpha brasses are often straight alloys of copper and zinc, other alloying metals including aluminium, iron, tin and manganese are added to alpha/beta brasses, and their strengthening effects are in the order in which they are named, that of aluminium being greatest.

Very often lead is introduced into brass to improve its machinability. When such a 'free-cutting' brass is machined on a lathe, the metal, which contains particles of lead, does not cling to the cutting tool in long spirals but breaks off in small chips, so a leaded brass can be machined at a much higher speed than would be possible in the absence of lead. To maximize the cutting speed, about 2.5 to 4.5 per cent lead is included in the alloy, which can then be used for making screw-threaded products. The presence of such an amount of lead in the brass brings about some deterioration in the mechanical properties and tends to make hot stamping difficult, so if brass has to be shaped by hot stamping and then machined, only about 1 to 2.5 per cent of lead is introduced.

BRONZE

Strictly speaking, bronze is an alloy of copper with tin, but the word has come to signify a 'superior' material, as compared with 'common' brass. For example, silicon bronze and aluminium bronze contain no tin; 'manganese bronze' is a high tensile brass, containing neither tin nor manganese.

The famous Ghiberti doors at Florence and *The Thinker* have been described on pages 167 and 70, but tribute must also be paid to the Marinelli family who, for over a thousand years, have been casting bells in the Italian hill-town of Agnone. A picture of the Perestroika Bell presented by Pope John Paul II to Mikhail Gorbachev is shown in *Plate 22*. The Marinelli concern, which supplies large and small bells to many countries, uses a bronze containing 87 per cent copper, 22 per cent tin and small amounts of other elements. A construction of bricks is prepared which corresponds to the inside shape of the bell. This is covered with clay to produce a smooth surface onto which wax casts of the decorations and inscriptions are moulded, then an outer mould of clay is spread to the required thickness. The complete mould is fired so that the wax melts away, leaving a cavity into which molten bronze is poured. When the casting has cooled, the outer clay covering is broken away, and the bell is removed and trimmed ready for testing and delivery.

The sonority of bronze is also shown, in a less dignified manner, in the vast range of bronze cymbals available to symphony orchestras, pop and rock groups, and marching bands. Depending on the tone required, the bronze can contain between 15 and 20 per cent tin. Starting from sheet metal where the grain structure is controlled so as to give a focused sound, the discs are shaped and hammered into the final product. The processes used to meet the very demanding standards of percussionists have for many generations been the monopoly of a family of Turkish craftsmen, now operating in America and Britain.

A modification of copper–tin bronze is phosphor bronze. One type of this alloy contains 4.5 to 6 per cent tin and less than 0.3 per cent phosphorus, which then exists in solid solution in the bronze. Such a phosphor bronze is very suitable for springs and electrical contact mechanisms where the alloy's resilience, non-magnetic properties and freedom from corrosion maintain regular working. When over 0.3 per cent phosphorus is present, the surplus separates as a hard constituent, Cu_3P; this type of phosphor bronze, which usually contains about 10 per cent tin, is extensively used in the form of castings. It is harder than the first type and makes a good bearing material, and is also used for components which endure heavy compressive loads, such as parts of moving bridges and turntables, and rolling-mill bearings.

The alloys known as gunmetals are copper-based alloys containing tin; this type of alloy was known from early times, when it was used for cannons. Until 1942 the Victoria Cross, Britain's highest award for valour, was struck from the metal of Russian guns captured at Sebastopol during the Crimean War; nowadays they are made by a London foundry. Because gunmetal can be cast with ease it still enjoys a wide reputation, though not for making guns. One of the traditional alloys, called Admiralty gunmetal, contains 88 per cent copper, 10 per cent tin and 2 per cent zinc, and has been used for marine purposes, for pump bodies and in high pressure steam plants. The leaded gunmetals, headed by the ubiquitous 85–5–5–5 alloy (85 per cent copper with 5 per cent each of tin, zinc and lead), dominate the sand-founding industry for general purpose and pressure-tight castings.

LEADED BRONZE BEARINGS

The function of lead in a bearing bronze is to act as a sort of metallic lubricant when the oil film breaks down. In some bearings the leaded bronze consists of copper with up to 30 per cent lead, with additions of 5 per cent or more of other elements such as zinc, tin or nickel. The copper provides high thermal conductivity, which assists in avoiding overheating, while the minor additions improve the lead distribution and increase the mechanical strength of the bearing. These alloys are sometimes used for grinding-machine bearings; the lead acts as an absorbent, in which minute particles of grit can safely embed themselves and thereby reduce wear on the remainder of the bearing surface.

ALUMINIUM BRONZE

The aluminium bronzes, previously mentioned on page 41, are copper–aluminium alloys to which other elements such as iron and nickel can be added. They have an attractive colour and a strength comparable with that of steel, and they resist corrosion well. Products requiring cold working, such as tube and sheet, are generally made in alloys with less than 8 per cent aluminium; they are softer and more ductile than the alloys with between 8 and 11 per cent aluminium, which are used in the sand-cast, permanent mould cast and hot-worked forms. Typical examples are gear selector forks for automobiles and components of pleasure cruisers. The propellers for the *Queen Elizabeth 2* are made of a nickel–aluminium bronze. An alloy containing 9 per cent aluminium, 4.5 per cent nickel and 5 per cent iron is used for the propellers of large crude oil carriers (described on page 69 and illustrated in *Plate 9*).

The strength and corrosion resistance of aluminium bronze are demonstrated by its use for pumps, valves and pipes on oil and gas platforms. Sea water is employed for cooling and firefighting, and is pumped up 30 metres or more from sea level. The aluminium bronze pipes are centrifugally cast in metre lengths, and the sections welded together. Aluminium bronze is also used in the gear wheels for braking systems on railways.

OTHER ALLOYS

Although copper has high electrical conductivity it is possible, by alloying, to decrease the conductivity so much that some copper alloys have great resistance to the passage of electric current. One such copper alloy contains 13 per cent manganese and 2 per cent aluminium. These alloys also have the property that their electrical resistance does not change with variation in temperature; they are used for underfloor heating and for resistances to control the speed of electric motors.

Other copper alloys are referred to elsewhere: copper–nickel alloys and 'nickel silvers' on page 193 and beryllium copper on page 202. Partly because copper forms alloys with so many other metals, but also because it has been known for thousands of years, very many copper alloys have been in use, ranging from the modern temper-hardening copper–chromium alloys to the tin bronzes, which are as old as the art of metallurgy.

1. Aerial view of the Bingham Canyon copper mine – the world's biggest quarry

2. The blast furnace at Redcar (*Courtesy British Steel plc*)

3. The control room of the Redcar blast furnace (*Courtesy British Steel plc*)

4. The 'Diane de Gabies', an aluminium statuette cast in the mid-nineteenth century (*Courtesy The Aluminium Federation*)

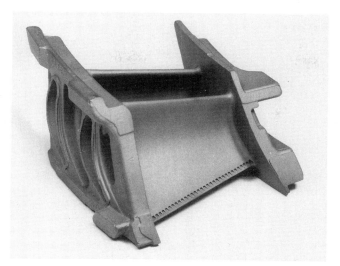

5a. A second-stage vane segment for the Pratt & Whitney
F100-PW-200 fighter engine (*Courtesy Howmet Corporation*)

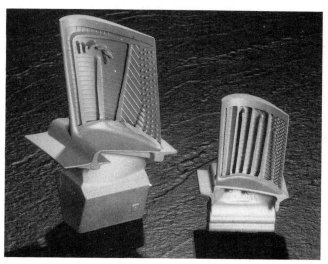

5b. Cutaways of blades for two jet engines: (left) single-crystal
casting for the Pratt & Whitney PW 2037 engine, and (right)
a directionally solidified component for the General Electric
Company's F 404 engine (*Courtesy Howmet Corporation*)

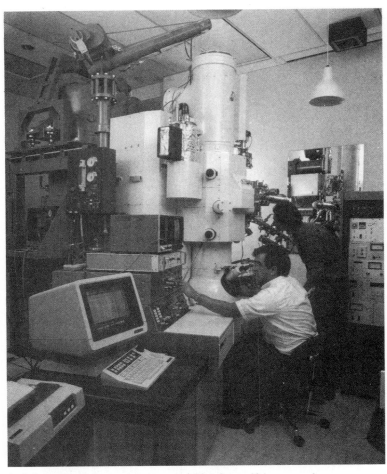

6. A Japanese electron microscope capable of magnifying more than a million times (*Courtesy Jeol (UK) Ltd and Xerox, Palo Alto*)

7. A scanning Auger microscope (*Courtesy Materials Research Laboratories, Department of Defence, Government of Australia*)

8. The bubble raft experiment

9. Nickel–aluminium bronze propeller for the oil carrier STS *Hellespont Paramount* (*Courtesy Stone Manganese Marine Ltd*)

10. Casting a marine propeller: the mould in the course of construction (*Courtesy Stone Manganese Marine Ltd*)

11. The largest press of its kind in the UK: a 6000 tonne mechanical press, capable of producing forgings up to 60 kg in weight (*Courtesy Garringtons Plant, a member of United Engineering and Forging*, UK)

12. A 4-high mill for cold rolling aluminium strip 2 metres wide, in operation at Commonwealth Aluminum Corporation, USA (*Courtesy Davy McKee, Poole, UK*)

13. A robot stacking bags of coin (*Courtesy the Royal Mint*)

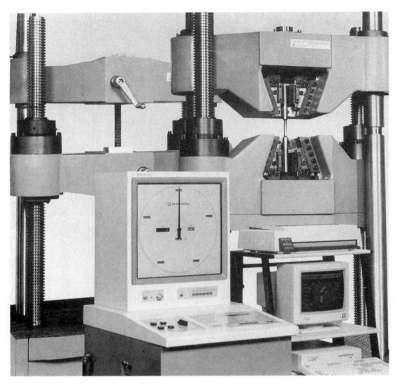

14. A universal testing machine for measuring the tensile and compressive strengths of metals (*Courtesy Shimadzu Corporation*)

15. The world's first iron bridge, over the River Severn at a place now called Ironbridge

16. The SS *Great Britain*, now at Great Western Dock, Bristol
(*Courtesy SS Great Britain Project*)

17. Continuous casting of steel: British Steel's eight-strand machine at Lackenby (*Courtesy Davy International*)

18. Jacket section of a North Sea oil production platform under construction at Highlands Fabricators, Scotland; the double-decker bus parked alongside gives an idea of its size (*Courtesy British Petroleum plc*)

19. Ipsen Abar and ERCO vacuum heat treatment furnaces, with capacities of 2¼ tonnes and a maximum operating temperature of 1350°C
(*Courtesy Bodycote (UK) Ltd*)

20. 'Eros' – a classic example of aluminium casting

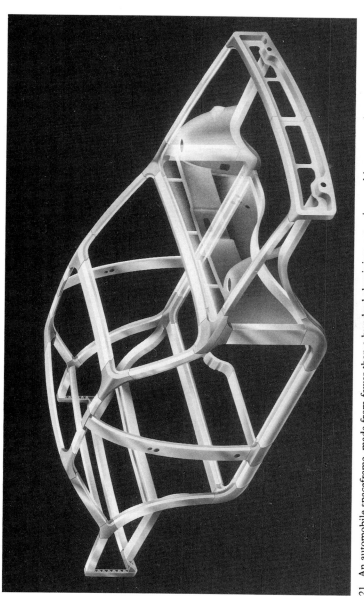

21. An automobile spaceframe, made from fewer than a hundred aluminium extrusions and interconnecting aluminium diecast nodes, robotically welded to the car body (*Courtesy Alcoa and Audi*)

22. Two views of the Perestroika Bell (*Courtesy Fonderia Marinelli*)

23. A Rolls-Royce RB 211 three-shaft turbofan (*Courtesy Rolls-Royce plc*)

24. A lightweight mountain bike with a diecast magnesium alloy frame (*Courtesy Kirk Precision Ltd*)

25. Consumable electrode vacuum arc furnaces for melting titanium at IMI Titanium's plant in Birmingham (*Courtesy IMI Titanium*)

26. Drilling machine in the 'crossover cavern' of the Channel Tunnel
(*Courtesy QA Photos Ltd*)

27. The Thames Barrier (*Courtesy A. D. Mercer*)

28a. A drilling bit with tungsten carbide teeth (*Courtesy Hughes Tool Company*)

28b. Robotic arc welding (*Courtesy ESAB Group (UK) Ltd*)

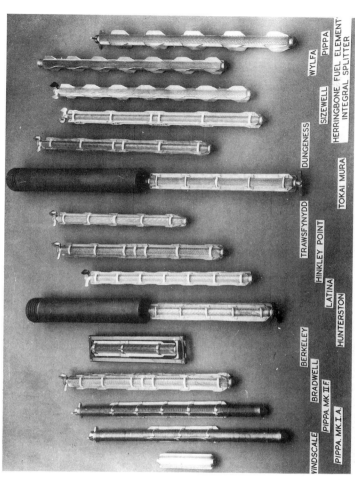

29. Fuel elements for various nuclear power stations
(*Courtesy United Kingdom Atomic Energy Authority*)

30a. Surface of cast antimony, showing dendritic structure

30b. Pearlite: a steel with about 0.9 per cent carbon

30c. Martensite: the structure of a hardened steel

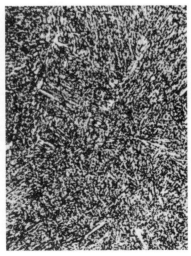

30d. The structure of a steel which has been hardened and tempered

31a. Cuboids of antimony–tin intermetallic in a tin–antimony–lead bearing alloy

31b. Alloy of 60 per cent copper and 40 per cent silver, showing dendrites

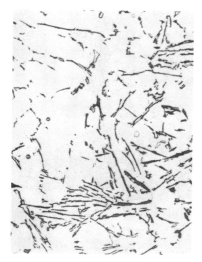

31c. Alloy containing 88.5 per cent aluminium and 11.5 per cent silicon (unmodified)

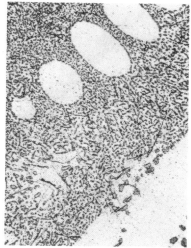

31d. The same alloy as in Plate 31c, modified by the addition of 0.05 per cent sodium

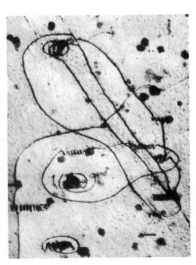

32a. Dislocations in aluminium which has been cold worked

32b. Dislocation lines from particles of aluminium oxide in brass

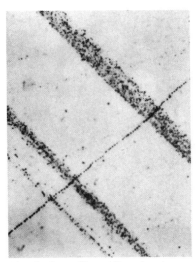

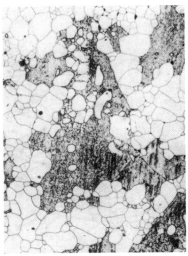

32c. An iron–silicon alloy, showing slip bands

32d. The same iron–silicon alloy as in Plate 32c after annealing, showing dislocations and recrystallization

18

THREE COMMON METALS: TIN, ZINC AND LEAD

TIN

Tin is a soft, weak metal, slightly less dense than iron and with a low melting point (232°C). Normally it costs around £4000 ($6000) per tonne, but in some years the price has fluctuated wildly. It rose from £3000 per tonne in 1975 to £10,000 per tonne in 1985, and returned to about £4000 per tonne in the early 1990s. The high price of tin means that it is used mainly where maximum advantage can be taken of its corrosion resistance and non-toxicity: from cars to cans, and from the manufacture of plate glass to the assembly of micro-circuits. Yet most people probably think tin is cheap: one often hears, 'It's only a bit of tin.' This reflects the popular confusion of tin with tinplate. Tinplate is thin steel sheet coated with a much thinner layer of tin. It is an excellent combination of metals because it has the strength and rigidity of steel, plus the attractive appearance, corrosion resistance and good weldability of the tin coating. If steel alone were used for food storage the metal would rust, while tin alone would be too soft and much too expensive.

The story of tinplate's progress from an interesting invention to a mass-production industry began with Napoleon Bonaparte. Napoleon was concerned with the problem of feeding his troops who, as everyone knows, marched on their stomachs and whose efficient movement over great distances in good fighting order provided one of the shock tactics which enabled Napoleon to win so many battles. In 1795 he offered a prize of 12,000 francs for a method of keeping food fresh for long periods. The winner was Nicholas Appert of Paris, who discovered that when foodstuffs were boiled in glass bottles and immediately sealed, they would keep for several months. This led to other processes being tried. Among them was a method patented by an Englishman, Peter Durand, of enclosing food in containers of iron coated on the inside with a thin film of tin.

In the nineteenth century the manufacture of tinplate was practically a British monopoly, until 1891, when the USA imposed tariff protection to its manufacture. By 1912 the UK and the USA were making three-quarters of a million tons each, but during the First World War British production was halved while that of the USA doubled. In the meantime, great strides had been made in the rapid production of cans, starting in 1846 when a device was invented for increasing the rate of output from six to sixty per hour. The first automatic can-making machinery was introduced during the 1880s, leading to the modern plants which produce a thousand cans per minute.

Tinplate is made with mild steel containing about 0.1 per cent carbon. The steel is continuously cast and hot rolled into strips about a metre wide, 3 mm thick and up to 1000 m long. Coils of this steel, weighing up to 12 tonnes, are brought from the hot rolling mill, pickled in acid to remove scale, and prepared for cold rolling in which the thickness is reduced from 3 mm to the finished gauge, which nowadays can be as little as 0.12 mm. The cold rolling is done by a 5-stand high speed tandem mill (page 83). The maximum speed at exit is about 1500 m per minute. After all traces of rolling lubricant have been removed, the steel is annealed by heating in a non-oxidizing atmosphere. Then it is given a final, very light rolling treatment without lubrication, which confers the required mechanical properties and surface finish.

Over 95 per cent of world production of tinplate is manufactured by electrodeposition in continuous automatic plants with annual capacities of up to 200 000 tonnes of finished tinplate, able to handle the strip at speeds of over 300 m per minute. Electrodeposition has almost entirely superseded the old hot-dip process of immersing sheets in molten tin because of its advantages in production speed, control of thickness and uniformity of the coating, and its ability to produce a thinner deposit. It also has the special advantage that different thicknesses of tin may be applied to the two faces of the strip. The coating thicknesses in common use are in the range 0.0004 to 0.002 mm.

In about 60 per cent of tinplate production, the tin is electrodeposited from a solution containing stannous sulphate, phenolsulphonic acid and organic addition agents which ensure a smooth coherent coating. The process is often called 'Ferrostan'. The steel strip is fed continuously through what is aptly called a 'vertical serpentine' electrolytic tinplate line. It is about 100 m long, and at any one time a length of over 900 m of strip is travelling up and down in vertical loops as it is uncoiled, cleaned, pickled, plated, momentarily heated to fuse the tin, passed through a 'passivation solution', oiled, automatically inspected and re-coiled. At full speed the tinplate takes only about three minutes to pass from one end of the line to the other. The brief fusion of the tin coating, known as 'flow brightening', gives it its brilliance. The passivation solution, of sodium dichromate, confers improved resistance to oxidation and staining; the oiling process is applied to ease the handling of the tinplate in can-making.

Most of the remainder of the electrolytic tinplate production uses what is called the 'halogen process' in which the tin is deposited from a weak acid solution containing stannous chloride and the fluoride of an alkali metal with an organic addition agent, and the steel passes through the cleaning and plating sections horizontally. One face is plated before the strip is looped back for plating the other face; otherwise the sequence of operations is as described for the vertical serpentine process.

Tin's low melting point and ready ability to alloy with other metals are reasons for its good solderability, so tin and tin–lead alloys are frequently used to coat electrical or mechanical components which are later required to be soldered. A large tonnage is required for electronic and electrical uses. The rapid developments in these industries have made it necessary to devise fast, automatic methods of soldering. Printed circuits are produced by wave soldering, in which the circuit assembly is passed across the crest of a wave of continuously circulating molten solder. In the past the seams of tin cans were automatically soldered, but today they are produced either by welding the side-seam instead of soldering, or by drawing processes which produce a container with a seamless body and one integral end.

Pewter is a tin alloy: there are two compositions in general use, one with 6 per cent antimony and 2 per cent copper; the other contains 4 per cent antimony and 2 per cent copper. The alloy is gravity diecast in Germany to make some very splendid beer tankards. Candlesticks, tea and coffee serving trays, and salt and pepper cruets are among the many other items of pewter available. Although pewter may be regarded as an old-fashioned metal, it is becoming increasingly popular with the introduction of new methods of manufacturing pewterware, and under the stimulus of modern design. Pewter is an easy metal to fashion and many people interested in crafts have had their first experience in metal-beating with pewter for ornaments and vessels.

Some bearing metals contain tin: for example, one bearing metal used in diesel engines and generators contains 90 per cent tin, 7 per cent antimony and 3 per cent copper. At the other end of the scale some lead alloy bearing metals contain 5 to 10 per cent tin. An important bearing alloy for automobile engines is an 80 per cent aluminium, 20 per cent tin alloy developed in Britain and now used throughout the world. Bronze and gunmetal, two well-known alloys containing tin, were discussed in the previous chapter.

A fairly recent development illustrates one of the most exciting uses of tin. In the Pilkington float glass process, molten glass is run onto a bath of pure liquid tin; the contents of the bath amount to anything up to 100 tonnes. Tin was chosen for this purpose because of its combination of properties: the requirements were for a supporting liquid which would have a density greater than glass, be molten below 600°C, but have a high boiling point and have virtually no chemical interaction with glass.

ZINC

The campfire of our forefathers led to the discovery of many metals. When oxide ores of copper, tin, lead or iron were fortuitously heated below a fire, the carbon in the burning wood united with the oxygen in the metal compound, leaving a more or less pure metal to be found when the fire had died down. Zinc, however, could not have been discovered in this way; at bright red heat the liquid metal boils and so, although metallic zinc may have been formed in the campfire, it would have turned to vapour which quickly oxidized in contact with the air, forming a cloud of white fumes, easily mistaken for smoke.

Long before zinc was known as a metal, the Romans mixed calamine, which contains zinc carbonate, with copper ores; the smelting of the two materials produced brass. Zinc was not isolated for many hundreds of years after the discovery of brass; it was first made in Sumatra and China, whence it was exported to Europe in the early seventeenth century. England was the first European country to develop its manufacture; William Champion of Bristol was smelting zinc on a commercial scale in 1738.

The metal is contained in zinc sulphide ore deposits widely distributed throughout the world. Important mines are in Canada, Australia, Peru and the USA. Zinc is often smelted or electrolytically refined in the countries where the mines are situated, but large quantities of ore, usually concentrated by the flotation process (page 15) are also sent for treatment to Europe, Japan and the USA, whose requirements for zinc metal exceed the capacity of their mines to supply ore. Currently, the world's annual consumption is about $7\frac{1}{2}$ million tonnes.

The original process for the production of zinc was a thermal one. Zinc oxide formed by roasting sulphide ore was heated to a temperature of about 1100°C with anthracite or a similar carbonaceous material in banks of small horizontal fireclay retorts. Zinc was formed as a vapour which was caught as liquid metal in condensers adjoining the retorts. In the late 1920s an improvement to the thermal process was developed. In this, the vertical retort process, a briquetted mixture of roasted concentrates and bituminous coal is heated in large vertical retorts made of silicon carbide bricks in which reduction to the metal can proceed continuously. The daily output of a vertical retort is about 9 tonnes of metal. By a subsequent process of fractional distillation, zinc of a purity greater than 99.99 per cent can be produced for the manufacture of zinc alloy diecastings and for other purposes where high purity is needed.

The electrolytic process, developed during the First World War, now accounts for about four-fifths of world zinc production. Roasted concentrates are leached in sulphuric acid and, after purification of the solution, zinc is deposited electrolytically on aluminium sheets from which it is stripped off, melted and cast into slabs. The acid is simultaneously regenerated and used repeatedly. The purity of the metal so formed is also about 99.99 per cent.

The Imperial Smelting process

After works experiments which started in the 1940s, the Imperial Smelting Corporation of Avonmouth announced in 1957 the successful development of a blast furnace for making zinc. This notable achievement involved the simultaneous smelting of zinc and lead, compounds of the two metals normally occurring together in nature.

In the Imperial Smelting process, zinc and lead sulphides are roasted to produce oxides which are charged in the blast furnace with suitable amounts of coke. Pre-heated blast air enters the furnace through water-cooled tuyères. Slag and molten lead containing precious metals and copper from the charge are tapped from the bottom of the furnace and separated. Zinc leaves the shaft as a vapour in the furnace gas containing carbon monoxide and carbon dioxide, and is shock cooled in a lead splash condenser, the zinc vapour being absorbed by the lead. Next molten lead containing zinc in solution is pumped out of the condenser and cooled so that zinc comes out of solution and floats on the lead. The zinc layer is poured off and cast, and the cooled lead is recycled to the condenser.

The invention of the lead splash condenser was the key to the process, as rapid cooling prevented the oxidation of the zinc vapour by carbon dioxide which had frustrated earlier researchers. There are now twelve

Imperial Smelting furnaces, together producing about 900 000 tonnes of zinc and 400 000 tonnes of lead each year. Some of the zinc is then redistilled to make high purity metal. It is quite extraordinary that lead, which is a completely undesirable impurity in zinc alloys, should play such an important part in this process and is then entirely removed from the zinc.

Uses of zinc

The galvanizing of steel, the manufacture of zinc alloy diecastings and the production of brass of together account for over 80 per cent of the total consumption of zinc. Its other uses include rolled zinc for building, and zinc powder for protective paints. Zinc oxide is used in the vulcanization of rubber, in ceramics, in pharmaceuticals and in the production of a wide range of zinc compounds including zinc sulphate, zinc chloride and zinc phosphate, all of which have large-scale industrial uses.

Zinc sheet was adopted as a roof covering in Europe early in the nineteenth century, particularly in France, Belgium, Germany and the Netherlands; more than 200 000 tonnes are required in these countries each year for roofs and cladding, which can last for a century. Uses of zinc in building in the UK are presently small but are growing as architects turn to longer-lasting, maintenance-free materials.

Zinc coating processes

The biggest single use of zinc is for coating on steel to protect it against rust and the most important process for doing this is hot dip galvanizing, in which steel articles are immersed in molten zinc. The many uses of galvanized steel range from household buckets to electric power transmission guards, highway guard rails and the underbody protection of automobiles. Over the last fifty years the large-scale production of continuously galvanized steel strip has replaced individually dipped sheets. This material has a ductile coating and will stand the same amount of deformation as the steel itself.

Electro-galvanizing consists of electroplating zinc on the previously cleaned article. Although electro-galvanized coatings have a good appearance, they are applied primarily to prevent corrosion. Zinc plating by a continuous process is used to protect steel strip. Special automated processes are used for small parts such as nuts and bolts. The coating is thinner than that applied by hot dipping, and requires additional protection, generally by paint or a plastic powder coating.

Zinc may be coated onto metals in several other ways. In one method zinc is sprayed from a 'metallization pistol'. Zinc wire is fed into the pistol, where it is melted by an oxy-gas flame or electric arc and, in atomized form, blown onto the article by compressed air; the action is similar to that of a scent spray. This process is generally used for steel structures too large to be hot dipped, or where the temperature required for galvanizing (about 450°C) would cause problems. Metal-sprayed coatings, particularly when they are subsequently painted, give very good protection. Aluminium is also metal-sprayed.

Sherardizing is another coating process. Here, steel articles are placed in a rotating drum together with zinc powder and fine sand. The container is then sealed and heated for several hours at about 375°C, at which temperature the zinc diffuses into the steel, giving a fine-grained protective coating of zinc–iron alloy. Small steel parts to be sherardized are cleaned and then tumbled in a drum containing zinc dust, tiny glass balls and an activating solution. The balls hammer the zinc onto the steel surfaces, to which it becomes cold welded. In this case no heating is needed, so the properties of the steel are unchanged.

Zinc alloy diecastings

The low melting point and high fluidity of zinc alloys, coupled with their high strength, have enabled zinc alloy diecastings to penetrate markets in many different industries. One of the first of these was car production, where the strength, accuracy and rapid output were ideally suited to mass-production techniques. Today, the automotive industry remains the biggest single market for zinc diecastings, including components of fuel systems, brakes, windscreen wipers and steering column locks. External parts are also made from zinc diecastings: door locks and handles, lamp surrounds and door mirrors. When necessary they are plated with a copper–nickel–chromium alloy or finished with the same paint system as the rest of the vehicle. Other examples include builders' hardware, small power tools, electrical and electronic components, domestic appliances and audio-visual equipment. Components range in weight from several kilograms to a few milligrams – as far as we can ascertain, the smallest zinc alloy diecasting is a wedge pin for watches, weighing only one three-hundredth of a gram.

Pure zinc is unsuitable for diecasting: the parts would have poor strength, and unalloyed zinc dissolves iron and steel and thus would 'galvanize' the die. Alloys of zinc with 4 per cent aluminium and small quantities of magnesium, first produced in the 1930s and still used in very large tonnages today, do not react readily with the steel of dies and have the characteristic strength and other properties which have made the zinc diecasting industry flourish. These alloys are frequently referred to as 'Mazak', a name derived from the initial letters of 'magnesium, aluminium, zinc and kopper'*; in the USA they are known as 'Zamak'. Both these names may be applied to the products of individual companies.

Within the last decade the original zinc alloys with 4 per cent aluminium have been joined by others containing 8, 12 and 27 per cent aluminium. This extends the range of application of zinc diecastings because the new alloys have improved properties, particularly at temperatures above 100°C. All zinc casting alloys are sensitive to the effects of traces of certain impurities. This is in contrast to many other foundry alloys, for which small amounts of impurities are far from being catastrophic. Zinc castings contaminated with minute traces of lead, tin and cadmium are brittle and subject to intense intergranular corrosion. For this reason, 'special high grade' zinc (minimum 99.995 per cent pure) has to be used in making the alloy; indeed, the introduction of this grade of zinc in the 1930s was an essential factor in the establishment of the zinc alloy diecasting industry.

The ability to maintain close analytical control on alloys and castings led in the 1950s to the establishment of certification schemes for zinc diecasters in several countries. These schemes ensured that they carried out all the necessary good housekeeping and sufficient routine analyses, with spot checks from independent outside bodies. Today the demands for quality assurance from all users of zinc alloy castings have led to these requirements being incorporated in more general schemes of quality management.

LEAD

The Romans' elaborate system of water distribution made it necessary to obtain large amounts of lead; Greece, Spain and Britain were the principal sources of supply. For lining Roman baths the metal was cast into thick sheets. Water-pipes were made by bending lead sheets over a rod so that a tube shape was produced, the joint being sealed with solder or with lead by the Roman *plumbarii* (lead-workers). Some of the lead pipes and conduits made two thousand years ago can be seen in museums

---

* In the early days of the industry about 3 per cent copper was included; now zinc diecasting alloys are either copper-free or contain only 0.75 per cent copper.

in Rome and in such buildings as the house of Livia, wife of Augustus Caesar, on the Palatine Hill. Others have been discovered at Pompeii and Bath.

A property which makes lead so easy to work is its low melting point (327°C). Its ease of casting is one of the reasons why for centuries lead was associated with the manufacture of printers' type metal; all editions of this book until 1980 were printed using an alloy consisting of 85 per cent lead, 11 per cent antimony and 4 per cent tin. (The present edition has been produced by computerized photo-typesetting.)

The world's annual production of lead is about 6.5 million tonnes, though less than half of this is newly smelted metal: a large amount is recycled from scrap batteries, sheet, pipes and cable sheathing. Oxides and impurities are removed, the alloy is brought to the required composition, pumped from the refining furnace and automatically cast into ingots. Apart from scrap reclamation, the main sources are lead ores in Australia, the USA, Canada and China.

About 100 000 tonnes of lead, as metal and oxide, is used each year for making car batteries in Britain, while in the USA over a million tonnes, representing about 80 per cent of their total consumption of lead, is used for the same purpose. The so-called automobile SLI battery (starting, lighting and ignition) is made with flat lead alloy positive and negative grids. These are coated with a lead oxide/sulphuric acid paste, and cured and electrically formed to give the desired properties. Until recently, lead alloys with 3 to 5 per cent antimony were used, but it was found that lowering the antimony content reduced water loss from the battery and thus minimized the need for 'topping up' the electrolyte. Some manufacturers use grids made from lead with about 0.09 per cent calcium, which reduces water loss to such an extent that the battery can be sealed and becomes maintenance-free. Normally 0.3 to 0.5 per cent tin is added to the positive grids in these alloys to enhance their corrosion resistance. In Europe a hybrid design with a low antimony positive grid and a lead–calcium negative grid is popular. Although not a sealed battery, water loss is so low as to make it effectively maintenance-free as far as the owner of the automobile is concerned. The manufacture of battery grids is a highly mechanized process. A common method has been to use high speed gravity diecasting; in a more recent development grids are produced in strip form by a continuous casting process at a rate of 400 a minute.

Advances have been made not only in the production methods but also in the performance of SLI batteries, particularly in relation to their power-to-weight ratio and their ability to give better 'cranking power' to start engines in difficult cold weather conditions. In fact, the electric power per unit weight has doubled in the last ten years. Such improvements increase the potential of urban electric automobiles or 'hybrids' in which electric batteries and internal combustion engines are combined to reduce city-centre pollution.

The future development of the electric automobile has been given a stimulus by Californian legislation which aims to reduce air pollution in the Los Angeles area. The regulation stipulates that by 1998 any large automobile manufacturer must have at least 2 per cent of its Californian sales as 'zero emission' vehicles. Fuel cell and battery powered units are being actively researched; the percentage is set to increase to 5 per cent by 2001 and 10 per cent by 2003. This may seem a drop in the ocean, but it represents some progress and, of course, further developments in the efficiency of batteries will lead to more electric cars. General Motors has already announced the Impact, based on a lead–acid battery system, which will undergo extensive road tests in 1994.

Research continues to develop other battery systems with increased power-to-weight ratios, but with most there are problems to be overcome in what is now a tight timescale if they are to satisfy the Californian regulation. Systems being looked at as alternatives to lead–acid include nickel/cadmium, nickel/nickel hydride, aluminium/lithium and sodium/sulphur. Many of the prototype cars being built are still based on lead–acid batteries as being the most reliable system, and work continues to improve the power-to-weight ratio and reduce recharging time.

For heavy duty applications such as in fork lift trucks for materials handling, in mining, on the railways or in submarines, different designs of battery are used. These can either be made from much thicker flat plate grids, or alternatively the positive grid can be made from lead–antimony alloy spines which are located in gauntlets filled with lead oxide, giving long life under arduous conditions.

Disabled people rely on electric wheelchairs and invalid carriages, while in hospitals, railway stations and airports electric vehicles are used to transport food, mailbags, luggage and people. A fast growing use for lead–acid batteries is in the provision of emergency power in places where a power cut would bring disaster. Common applications of such systems are in hospitals, computer installations, telephone exchanges and alarm systems. The fast growth of the market has been helped by the introduction of a new design – the valve-regulated battery. In such batteries the lead–acid chemistry is carefully controlled, so that any oxygen and hydrogen formed within the battery during use is recombined to form water so that the battery can be sealed and requires no maintenance. These batteries are used with systems that keep them fully charged and ready for use should an emergency occur. The growth in demand for cordless tools is also likely to present opportunities for small sealed lead–acid batteries. At the opposite end of the scale, very large battery installations have been built for levelling out the variations in demand on power utilities. One such installation in California contains 1000 tonnes of lead in batteries for storing power at times of low demand, to be released when the load peaks.

Lead has been used for buildings since early Egyptian times. In Britain there are many lead roofs which, like those on Westminster Abbey and St Paul's Cathedral, have lasted for several centuries. Those who admire the splendid pictures in stained glass windows do not always realize that lead 'cames' hold together the pieces of glass. Lead sheet is widely used for flashings and weatherings under windows, around chimneys and on parapet walls. Lead sheet can be bonded to plywood and chipboard, and such laminates make effective partitions, with good sound insulation. In the 'high-tech' area, leaded glass is used in computer screens and in TV tubes, to protect the viewer from the cathode rays generated in making the picture.

Lead will not dissolve in some liquid metals; for example, lead and aluminium will not mix. However, lead forms useful alloys with metals of similarly low melting points. Until recently, large amounts of lead–tin solders were required for plumbing and for assembling metal components, but now they are being phased out from domestic plumbing in favour of lead-free alternatives. However, lead–tin solders continue to be required in the fast-growing electronics industry.

In the past, lead has been thought of as somewhat unglamorous, though capable of doing a great deal of metallurgical donkey-work. However (to mix metaphors), lead, like other dark horses, often brings surprises. Many modern developments of lead and its alloys have been enterprising and sophisticated. In nuclear research establishments and in hospitals where X-rays, radium and radioactive isotopes are used in examination and treatment, lead alloyed with 4 per cent antimony is used in the form of interlocking bricks to form screens. Radiography and hospital treatment rooms are constructed with lead sheet attached to other materials such as plywood. Lead oxide can be incorporated in glass to make radiation-proof viewing windows in nuclear power stations. On a smaller scale, radioactive material is conveyed from one hospital to another in small containers, cast or machined from lead. An unusual application exploiting the property of radiation shielding is a lead alloy plate, 3 mm thick, with an irregular pattern of small grooves cast into the surface. These plates are designed to fit behind the keyholes of high security locks, to scatter away any radiation falling on them, thus preventing high-tech burglars from using radiography to make accurate images of the locking mechanism.

Lead is also used as a protective sheath on electric power cables. Although there has been substitution by plastics, lead is still the best material for underwater protection. Its corrosion resistance and density make it ideal for many applications, from tiny castings for precision instruments to multi-tonne yacht keels. Lead compounds such as white lead and red lead were commonly used in paints because of their excellent weathering or anti-corrosion properties. Also, lead-based compounds such as lead naphthenate were used to promote drying. Concerns over the toxicity of lead have resulted in the virtual elimination of lead compounds from domestic paint, and the white lead has been replaced by titanium dioxide. About a third of the metal's consumption is for chemical compounds of which the major part is as lead oxide for batteries, but to a smaller yet significant extent in glass, ceramics and stabilizers for plastics.

The tower at Pisa began to lean soon after its construction in 1173, and has been leaning at an increasing rate ever since. Now it is $5\frac{1}{2}$ degrees off perpendicular – the top overhangs the base by 4 metres – and it has sunk 3 metres into the soft soil on its south side. In 1989, on safety grounds, the authorities ordered its closure to tourists.

Now the tower is being stabilized with the help of lead. The process began with the placing of a number of 8 tonne lead counterweights on the north side. By the end of 1993, 600 tonnes of lead had been placed at the foot of the tower, reversing the tilt by just 3 centimetres. In 1995 the lead weights will be taken away; soil will be gradually removed from beneath the foundations, and water extracted from the soft, waterclogged clay that caused the tower to start leaning in the first place.

19

NICKEL

The name of this metal derives from the experiences of fifteenth-century miners who suffered from arsenic poisoning while attempting to recover copper from ores in the hills of Saxony. They assumed that their misfortunes and their inability to extract good enough copper were caused by evil spirits, so they named the ore *Kupfer-nickel*, 'Devil's copper'. The nature of the ore was explained by the Swede Axel Fredrik Cronstedt in 1751: it contained a compound of arsenic with a metallic element, which he named 'nickel'. The new metal appeared to be brittle and of no practical use, until fifty years later it was shown that the brittleness was caused by carbon and sulphur, which could be eliminated by additions of manganese and magnesium. Thus purified, nickel was found to be a strong and workable metal with good corrosion resistance.

In 1867 large deposits of nickel ore were found in New Caledonia, an island about 800 miles east of Australia. In 1883, during the construction of the Canadian Pacific Railway, a massive outcrop of nickel–copper ore was found near Sudbury; it remains the largest source of nickel in the western world. In addition, nickel ores come from the countries of the former USSR, Australia, Cuba, the Philippines, Indonesia and several sources in the USA.

The two principal nickel ores are nickel sulphide, yielding about 1 per cent metal, in the hard rock formations of Canada and Australia, and what are known as lateritic ores of the iron–nickel oxide type, yielding about 4 per cent metal, found near the surface in countries in equatorial regions. The deep underground mining of sulphide ore is both capital- and labour-intensive, but because the minerals can be concentrated by flotation, the extraction and refining processes are comparatively low-cost; furthermore, some sulphide ores contain other valuable elements such as copper, cobalt, gold, silver and platinum metals, most of which are absent from lateritic ores.

The smelting and refining processes vary according to the type of ore. The sulphide ores are concentrated by flotation or by magnetic separation. This is followed by a series of pyrometallurgical processes that yield an impure form of nickel, as pellets or powder, which will then, if necessary, be treated by refining processes. Some ores are given a complex sequence of solvent extraction treatments, based on carefully exploiting the rate of reaction and selective solubility of nickel, copper, iron and cobalt in ammonia under controlled conditions of temperature, pressure and agitation. The end product of this route is a pure nickel powder which may then be compacted into manageable shapes.

Lateritic ores cannot be concentrated by froth flotation, so the ore is dried (some contain over 50 per cent moisture) and then smelted in blast furnaces to yield a suitable concentrate for further smelting, leaching and refining. After these preliminary treatments, the lateritic ores are either smelted to produce ferro-nickel for use in alloy steelmaking, or the processing is continued to refine the smelter concentrate and then convert it to pure nickel by a reduction process, or by electrolytic refining.

One of the most efficient methods of refining nickel is the carbonyl process, developed by Ludwig Mond and Carl Langer in the latter part of the nineteenth century. They discovered that freshly reduced nickel reacts with carbon monoxide at about 50°C to produce gaseous nickel carbonyl, which can then be decomposed into nickel and carbon monoxide by heat at about 200°C. This reversible process is the basis of automated plants. Nickel oxide, produced by roasting nickel sulphide, is reduced to an impure form of nickel by treatment with hydrogen. The metal is fed into a volatilization kiln in which it meets a counterflow of carbon monoxide at 60°C. Decomposition of the nickel carbonyl gas is effected on the surface of pre-heated nickel pellets which flow continuously through a reaction chamber until they grow to a marketable size.

The present world production of nickel amounts to about 850 000 tonnes a year; 60 per cent of this is required for stainless steels, 20 per cent for non-ferrous alloys and about 5 per cent for electroplating. The remainder is used in nickel alloy castings, catalysts and other applications.

USES OF PURE NICKEL

Nickel is a metal similar to iron in some of its properties. It has a melting point of 1455°C, nearly as high as that of iron; it has slightly lower strength and hardness and is magnetic, though to a lesser degree than iron. In contrast to iron, however, nickel is strongly resistant to corrosion;

for example, it is not corroded by alkalis or chlorine, and is used for the construction of plant making those materials.

The best-known use of pure nickel is in electroplating, where it forms the corrosion resistant layer underneath a thin chromium topcoat. This durable system of decorating and protecting base metals is widely used in office equipment, tubular furniture, domestic hardware, bathroom fitments, motor cycles, trucks and cars. After preliminary cleaning processes, articles to be plated are suspended in vats containing a solution of nickel sulphate, nickel chloride and boric acid, usually with small additions of organic chemicals to modify the deposit's appearance and properties. Strips of pure nickel, or titanium-mesh baskets containing nickel pieces, are also suspended; these are the anodes, and as nickel is deposited they dissolve and replenish the concentration of nickel in the electrolyte. Low voltage direct current flows from the anodes via the electrolyte to the articles being plated, which act as cathodes, and nickel is deposited on them. The contents of the vats are held at about 40°C and agitated, usually by air bubbling through the solution. Nowadays the whole sequence of operations is automated and computer controlled.

Modern plating solutions deposit nickel at high rates and provide a bright, smooth finish that needs little polishing. The micro-cracked chromium and microporous chromium topcoats now regularly applied to nickel plated articles have further enhanced the durability of this versatile surface finish. By making the surface of certain plastics electroconductive, nickel and chromium plating can also provide a hard-wearing finish to a variety of plastic moulded products.

Electroforming is another use for plated nickel in which a thick layer of the metal is deposited on a mandrel and then removed as a free-standing nickel object. The inside surface is a perfect replica in reverse of the finish and shape of the mandrel's surface, and nickel electroforms are used as moulding tools, printing plates, aerospace products, sieves, meshes and foils.

STAINLESS STEELS

These have been discussed on page 141. The martensitic and ferritic steels contain from 12 to 15 per cent chromium. The austenitic stainless steels containing nickel and chromium represent by far the largest use of nickel, which stabilizes the austenite and gives the steel added ductility, toughness and corrosion resistance. Recently, duplex stainless steels, with a mixed austenite–ferrite structure, have been developed.

NICKEL STEELS

In 1889, James Riley presented a now historic paper to the Iron and Steel Institute of Great Britain, giving a survey of the possibilities of nickel steels. Tests had shown that a steel containing 3 per cent nickel developed remarkable projectile-resisting qualities, coupled with tremendous toughness, making it suitable for armour plate. Riley's report, which shook the armament and engineering world, covered a wide range of steels with from 1 to 49 per cent nickel, and he announced that, by hardening and tempering, the almost unbelievable strength of 96 tons per in^2 ($1500 N/mm^2$) had been achieved. He confirmed that steels rich in nickel are practically non-corrodible, and he concluded, 'I find some difficulty in not becoming enthusiastic, for in the wide range of properties possessed by these alloys, it really seems as if any conceivable demand could be met and satisfied.' From that time onwards nickel alloy steels became vital materials, although reliable ways of producing even better properties in other alloy steels were later found. Most of them contain nickel, together with chromium and molybdenum. These alloy steels have now replaced 'straight' nickel steels in most of their uses.

CRYOGENIC STEELS

Modern processes which require liquid gases at sub-normal temperatures have increased the need for nickel in steels because it lowers the impact-transition temperature, at which steels become brittle. A relatively small addition of nickel to carbon steel can make an appreciable improvement but, for really low temperatures, down to $-196°C$ for containers of liquid nitrogen, steels containing 9 per cent nickel, which can be welded on site are used. For liquid helium, at a temperature of $-269°C$, austenitic stainless steels are required.

MARAGING STEELS

In the late 1950s a group of superalloy steels were developed, containing 18 per cent nickel, 8–12 per cent cobalt, 3–5 per cent molybdenum and small amounts of titanium and aluminium. They are called maraging steels because their structure is similar to that of martensite (discussed on page 135), and because they can be age hardened. These steels can be forged and machined in the unhardened condition and then brought up to a high tensile strength by a relatively low temperature heat treatment which does not distort or crack components of complex shape. Maraging steels are used for parts of aircraft undercarriages, rocket launcher tubes on submarines, portable military bridges and for complex tooling, such as that required in diecasting.

NICKEL CAST IRONS

Small additions of nickel to cast iron have been made for many years, but nowadays those with under 3 per cent nickel are not often specified. To quote one expert, 'Small amounts of nickel in cast iron are neither harmful nor beneficial.' The austenitic cast irons are of more importance. They contain at least 13 per cent nickel, with small additions of copper, chromium, manganese and silicon. These alloys have good heat and corrosion resistance and are used for body castings of large sea water pumps, power plant condensers and desalination plant. Martensitic cast irons, containing 4 per cent nickel with 2.5 per cent each of chromium and silicon, are intensely hard and are used for ball mill liners and for the bodies of pumps designed to handle abrasive slurries.

HEAT RESISTANT ALLOYS

A typical heat resistant alloy is based on 80 per cent nickel and 20 per cent chromium, plus small additions of rare earth metals to inhibit crystal growth at high temperature. Such an alloy is used for heating elements, ranging from the domestic electric toaster to giant electric furnaces. Like all heat resistant alloys, it resists destructive oxidation at high temperatures (e.g. 1000°C); it is ductile enough when cold to enable it to be manufactured into wire, rod, sheet, tube or strip which can then be fabricated by normal means and joined by welding.

Developments of this 80–20 alloy led to the range known as Nimonics. They are given serial numbers to indicate their suitability for operation at progressively higher temperatures. In Nimonic 75 the 80–20 binary alloy is strengthened by titanium carbide. Nimonic 90 is a nickel–chromium–cobalt alloy, to which small additions of titanium and aluminium are made. Other alloys in this range include Nimonic 105 and 115, containing molybdenum.

There are also several nickel–chromium–iron alloys, including Inconel, with about 77 per cent nickel, 16 per cent chromium and 7 per cent iron, used for furnace components and heat treatment equipment. Another heat resistant alloy is Incoloy, with 32 per cent nickel, 20 per cent chromium and the balance iron. Both Inconel and Incoloy can be modified by the addition of molybdenum, aluminium, niobium and some of the rare earth metals. They have excellent resistance to corrosion and good strength at normal and elevated temperatures.

SUPERALLOYS

Since the middle of the Second World War, the ever-increasing demands made on aero engines have been the motivation behind the search for new materials for service at high temperatures. These demands have been met by the development of superalloys, many of them based on nickel. Their largest use is in the gas turbine engine, which ingests air from the atmosphere, compresses it several times, then adds fuel and burns the mixture to produce turbine inlet gases in the temperature range 800–1600°C. Superalloys are used for components such as blades, vanes, discs and ducts.

There is a great variety of nickel-based superalloys; most of them contain 5 to 25 per cent chromium, up to 8 per cent aluminium, and small amounts of boron, zirconium and carbon. Some of them contain traces of one or more of cobalt, molybdenum, niobium, tungsten, hafnium and tantalum. In addition to aircraft, industrial and marine gas turbines, these alloys are used in nuclear reactors, power generators and space vehicles.

CORROSION RESISTANT ALLOYS

Nickel–chromium–iron and iron–nickel–chromium alloys feature in alloy systems where resistance to corrosion at high temperatures is the main objective, combined with strength at high temperature. A familiar application is the metal tube which encloses the heating element of electric cooker hobs. This alloy has to resist attack by numerous kinds of spilt cooking fluids as well as having to withstand mechanical damage by the cooking utensils. A more extreme application is the steam generator heat exchanger in pressurized water nuclear reactors. Other applications are in chemical plant furnaces, flame tubes and burners. Nickel–chromium–molybdenum alloys are used to contain highly corrosive chemicals.

LOW EXPANSION ALLOYS

The nickel–iron series of alloys have a low or even a zero coefficient of expansion; well-known tradenames are Invar and Nilo. One application is for clock pendulums, but a greater use is for bimetallic strips comprising two alloys of different coefficients of expansion, which bend when heated to operate temperature-controlled valves and switches. Another major use is for linings in ships' tanks for transporting liquid natural gas. The absence of expansion and contraction simplifies design. Some recently developed nickel–iron–cobalt alloys have a constant low coefficient of thermal expansion and high strength. They are used in gas turbine casings, shafts and shrouds. Their low coefficient of expansion enables close control of clearance and accuracy in turbine shafts.

NICKEL–COPPER ALLOYS

In 1905 it was discovered that a Canadian nickel–copper–iron ore could be smelted to produce an alloy with good corrosion resistance. The alloy, containing about 65 per cent nickel, 32 per cent copper and small amounts of iron and manganese, was tradenamed 'Monel' after Ambrose Monell, president of the International Nickel Company. It is resistant to sea water corrosion and has been selected to protect the steel legs of some oil platforms in the vulnerable splash zones above and below the tide-mark. A variation of Monel, containing aluminium, can be heat treated to give high strength as well as corrosion resistance, and is used for pump shafts and the propeller shafts of power boats. Monel is produced to this day, but by straightforward alloying. At the other end of the scale, the alloys rich in copper range from 90–10 to 60–40 compositions. The 90 per cent alloy features in ships' condensers and, in enormous quantities, in desalination plant for the Middle East. It has good resistance to corrosion by sea water and resists biofouling by marine organisms. This resistance to marine growth is important in applications such as power plant intake screens, fish-farm fencing and sheathing of offshore oilfield platforms. Another development is the cladding of ships' hulls to combat the higher consumption necessary to overcome the encumbrance by marine growth on the hull. A copper–nickel–iron–manganese alloy is used for condenser tubes that have to operate in the presence of sand-laden sea water.

Nickel silvers are a series of copper–nickel–zinc alloys with varying amounts of nickel, used for the manufacture of springs. The best-quality EPNS – electroplated nickel silver – tableware is based on an alloy containing 20 per cent nickel.

20

MAGNESIUM

Magnesium has been called the lightweight champion of metals. Aluminium is more than one and a half times as dense, iron and steel are four times as dense, and copper and nickel five times as dense. This low density, coupled with the relatively high strength of its alloys, has won magnesium its present place in industry. It is significant that magnesium ores, unlike those of some other metals, are so widely distributed over the Earth as to preclude the possibility of establishing a monopoly in them, but the metal itself is so difficult to extract that it is exploitable only by highly industrialized nations.

Of all the common metals there is none about which there is so little public knowledge as magnesium. This is because it is seldom employed for domestic articles such as kitchenware, so we rarely have a chance to handle it. We are familiar with iron, steel, aluminium, copper, tin, zinc, lead and nickel, and can usually distinguish one from another, but if we see a piece of magnesium it can easily be mistaken for aluminium. There is one important exception: magnesium is popularly associated with the production of intense light, as in fireworks, incendiary bombs and military flares. Consequently most people have difficulty in believing that such a metal can be used for structural and engineering purposes. The apparent contradiction is explained by the fact that, before magnesium can be made to burn, it must be melted and overheated while in contact with ample supplies of air.

The special methods required for smelting and refining magnesium and for casting and fabricating its alloys were pioneered in Germany, although valuable development work was done in Britain and the USA. There has been a remarkable increase in output over the last ninety years. In 1900, annual world production amounted to only 10 tons. This increased to 1000 tons in 1920, 20 000 in 1937, and about 240 000 in 1943 – an elevenfold increase in six years, due to the wartime needs of the aircraft industries. After the war the annual output fell considerably, but then rose steadily again, reaching about 300 000 tonnes in 1992.

Although pure magnesium has no great strength (about $100\,N/mm^2$), when suitably alloyed its strength can be trebled or even quadrupled. The alloying elements most widely used are aluminium and zinc to increase the mechanical properties, and manganese to improve corrosion resistance. Small amounts of silicon are included in many of the alloys. By far the most widely used casting alloy, specified as AZ 91 contains 8–9.5 per cent aluminium, 0.3–1.0 per cent zinc and about 0.2 per cent manganese. They are particularly suitable for diecasting.

At the other end of the scale, many important aluminium alloys contain magnesium; indeed, over half the total production of the metal is required for alloying with aluminium, and a considerable amount of this is for the immense tonnages of aluminium alloyed with small percentages of magnesium for the manufacture of soft-drink and beer cans. In addition, about 50 000 tonnes of magnesium is used each year in the desulphurization of steel.

In the search for alloy compositions suitable for aircraft, it was found that the addition of less than 1 per cent of zirconium had the effect of reducing in size the individual grains of which the alloy is composed, and thus increasing its strength. The grain size of a magnesium–aluminium alloy in the sand-cast condition is about 0.5 mm, but an alloy with 0.65 per cent zirconium has a grain size of only about 0.03 mm. Further research showed that the addition of thorium to magnesium alloys containing zirconium led to further improvements in their mechanical properties. Castings containing 0.7 per cent zirconium, 3.0 per cent thorium and 2.5 per cent zinc are used in jet engines. Another important range of alloys contains 0.7 per cent zirconium, 2–5 per cent zinc and 2–3 per cent cerium. A recent invention enables strong and ductile castings, for example the thrust reversers in the RB 211 jet engine to be made by heat treating these alloys in hydrogen, which modifies their microstructure. This is an unusual use of hydrogen, since in most metals it lowers mechanical properties.

Magnesium alloys can be subjected to all the usual metallurgical treatments, including sand-casting, diecasting, forging and pressing, but they are used mainly in the cast condition. No metal can be machined so easily or as fast as magnesium. This of course helps to reduce the cost of the final product.

The predominant use of magnesium has been in the automobile industry, led by Volkswagen. Up to 1980, 20 million of the famous 'Beetle' cars had been manufactured, containing about 400 000 tonnes of magnesium alloy diecastings, their air cooled engines being ideally suited for ultra-light alloys. The trend to produce front wheel drive, water cooled engines reduced Volkswagen's consumption of magnesium, but the total tonnage is still very large and the plants in Germany, Brazil and Mexico contain impressive examples of magnesium diecasting technology. The components include crankcases in the remaining air cooled engines, but their largest consumption of magnesium is represented by transmission housings, ranging in weight from 7 to 12 kg.

The light weight of magnesium is a great advantage in chain-saws, textile machinery, portable telephones, compact disc players and computers. Sporting uses include fishing reels, archery bows and 'Little League' baseball bats. One enterprising British manufacturer produces magnesium alloy diecast frames for mountain bikes and racing cycles that have done well in the Tour de France and other events (*Plate 24*).

It is interesting to observe that, because German manufacturers knew of the successful use of magnesium by Volkswagen, their chain-saw industry was quick to adopt the metal. In other European countries the acceptance of magnesium was slower, though now 'things are moving'. In the USA, more and more automobile components are being made in magnesium alloys; by 1992 over 10 000 tonnes of diecast parts were made. Furthermore, the use of magnesium for chain-saws and other lightweight equipment has developed rapidly there.

Potentially, magnesium could replace aluminium in many applications, the decision depending mainly on commercial considerations. However, the relative cost per unit weight per unit strength is not the only factor. Magnesium can be machined faster than aluminium. As a further bonus, magnesium alloys do not attack die steels in the same way that aluminium does. A diecasting die for magnesium will have a life of from three to five hundred thousand castings, compared with about half that number for aluminium alloys. This benefits American producers because their automobile components are needed in quantities of many hundreds of thousands, so the extension of die life is an added economic advantage.

Corrosion of magnesium, once looked upon as a major problem, was found to occur through the presence of iron, nickel and copper impurities. Now high purity magnesium alloys are available having excellent resistance to corrosion, and requiring little or no protective treatment. However, it is essential to ensure that the design does not introduce galvanic corrosion, as would be caused by, say, direct contact between a steel bolt and a magnesium casting in the presence of any liquid which could act as an electrolyte. A wide range of chemical coatings can be applied to magnesium components. Before painting it is usual to apply a chromate coating. If special anti-corrosion treatment is necessary, magnesium alloys can be protected by immersing the components in vats containing a hot solution of chromate salts. Black or golden oxide films containing chromic oxide are thereby deposited on the surface of the magnesium, and they form a protective layer which can act as a base for the application of paint for further protection. As an example of the corrosion resistance of modern magnesium alloys, Chrysler Corporation changed from aluminium to the high purity magnesium alloy AZ91D for the accessory drive brackets on its 1991 Jeep 2.5 and 4 litre engines; the components are not even painted.

Magnesium is the sixth most abundant element in the Earth's crust, occurring in such minerals as dolomite and magnesite. Brines contain magnesium chloride, and each cubic kilometre of sea water contains over a million tonnes of magnesium compounds. If it is difficult to visualize such a large volume, imagine 1½ cubic metres of sea water – that contains enough magnesium to make a bicycle frame.

Several electrochemical processes are used to extract the metal. In one system, crushed dolomite is roasted, mixed with sea water and fed into large tanks, where the insoluble magnesium hydroxide that has been formed settles to the bottom. It is heated to form magnesium oxide, which is mixed with coke and reacted with chlorine in shaft furnaces, producing molten magnesium chloride. This is electrolysed; chlorine given off at the carbon anodes is recycled, while the liberated magnesium floats to the surface.

In another process, brine containing about 10 per cent magnesium chloride is purified and concentrated, and the water is removed first by evaporation, and then by hot air and hydrochloric acid gas. The magnesium chloride is electrolysed into magnesium and chlorine. Dow Chemicals, the world's largest magnesium producer, claims to extract the raw materials entirely from sea water.

21

TITANIUM

Titanium ores are widely distributed; of the structural metals only aluminium, iron and magnesium are more abundant. The processes of extraction, melting and fabrication are all expensive because of the need to avoid contamination, mainly by oxygen and nitrogen, while the metal is molten. Titanium ores such as rutile (titanium dioxide) are first concentrated and then converted to liquid titanium tetrachloride, which is roasted for several days in an atmosphere of argon, with additions of either sodium or magnesium, producing titanium and readily removable sodium chloride or magnesium chloride. The spongy titanium metal produced by this process has a purity of about 99 per cent.

The great reactivity of molten titanium makes it impossible to melt in normal crucibles; a radically different melting technique had to be devised. The consumable-electrode vacuum arc furnaces shown in *Plate 25* use a direct current arc between an electrode made of the metal to be melted and the base of a copper crucible cooled by a liquid sodium–potassium alloy. The entire arrangement is enclosed in a vessel which either contains a non-reactive gas such as argon, or which may be completely evacuated. The electrodes are usually made of compressed blocks of titanium sponge; alloying additions are included at this stage. Normally a slender ingot is melted first, to act as the electrode for subsequent melting operations. In spite of their reactivity when molten, titanium and its alloys are cast into finished shapes for the aerospace and many other industries.

The combination of high strength, lightness and corrosion resistance accounts for the value of titanium and its alloys in modern technology. The performance of mountain and racing bicycles has been revolutionized by the use of titanium tube frames, and modern aircraft include an ever-increasing amount of titanium in their construction.

The alloys of titanium divide into several groups, depending on the amounts of alloying elements and the resulting crystal structures. Current alloys contain up to 25 per cent of added elements such as aluminium, vanadium, molybdenum, niobium, tin and zirconium. The alloys are readily fabricated into complex parts of high integrity for aircraft structures. One such alloy, containing 6 per cent aluminium and 4 per cent vanadium, exhibits superplasticity (the ability to be stretched by several hundred per cent without breaking, discussed on page 42). The technology of superplastic forming is now applied routinely for airframe components, including some in the new Rolls-Royce Trent engine. In many instances parts are joined during the same forming process by diffusion bonding, whereby the titanium alloy surfaces weld together by atomic diffusion, leaving virtually no trace of a joint-line.

Titanium alloys have been developed for components of aero engines capable of operating up to 600°C. Some titanium alloys can be heat treated to strengths of over $1400\,N/mm^2$, a magnificent combination of strength and lightness; they are used where maximum strength per unit weight coupled with fatigue resistance is essential.

Concorde, which flies at speeds of Mach 2, includes about 4 per cent of titanium alloys, mainly in the engine surrounds. The Boeing 777 includes 6 per cent of titanium in its structure. The European Airbus uses a wide range of titanium products, including wing access panels, landing gear brackets and many other components. Similarly, several modern helicopters rely on titanium alloys for rotor heads, blade attachments, anti-vibration mountings and weapon carriers. Current designs of military aircraft such as the Tornado contain about 25 per cent titanium, while the Lockheed Blackbird, capable of over Mach 3, is said to contain about 85 per cent of its structural weight in titanium alloys.

Plate 23 shows one of the RB 211 series turbofans, which has titanium alloy compressor discs, spacers, rings, casings and blades, some of which are the recently developed hollow 'widechord' blades. In the USA titanium alloys are used for airborne military equipment, such as mortar bases and armour plating. Outside the aerospace industry, titanium is being specified for many engineering purposes that call for a high strength-to-weight ratio. Typical applications include steam turbine blading, connecting rods, crankshafts, camshafts, outlet valves and springs in high performance engines.

Titanium is used in the production of chlorine because it is one of the few metals not corroded by the gas. Other functions in the chemical industry include the production of acetic acid, benzoic acid and ethylene amines. Recent developments of titanium to combat 'condenseritis' will be discussed on page 232. Because of its inertness to body fluids, titanium is now widely used for surgical implants. Titanium and its alloys can be drawn to 0.05 mm diameter wires for sutures used to join parts of bones after an accident or operation.

Titanium's light weight and resistance to corrosion have led to its choice for reinforcing the structure of the Parthenon in Greece and the salvaged remains of the *Mary Rose*. In 1545 this warship was waiting to engage the French fleet, when she sank suddenly, with the tragic loss of seven hundred men. Over four centuries later, the waterlogged remains of the ship were lifted to the surface and brought back to Portsmouth, where they are undergoing treatment at the Royal Naval Museum, not far from Nelson's *Victory*. Titanium was chosen to support the deck beams because of its light weight and resistance to corrosion. Forty-five adjustable vertical props and several horizontal ones were used. A constant spray of chilled water plays over the structure of the ship to prevent premature drying-out of the delicate timber, which will be sprayed with polyethylene glycol during the next ten years. Finally the structure will be dried under controlled conditions. The task should be completed by about 2014.

The life of high speed steel cutting tools is improved by coating them with a deposit of titanium carbide, or by depositing an underlayer of titanium carbide, followed by a series of gradations of titanium carbo-nitrides and finally a titanium nitride layer. This gradual change in composition is necessary to maintain the adhesion of the hard titanium nitride. Recent developments incorporate the alternate deposition of titanium nitride and alumina, totalling ten layers, each about 0.5 micrometres thick. Such products have increased tool life by as much as six times.

The amount of titanium ore that is mined is much greater than can be accounted for by the production of the metal and its alloys. By far the largest use is as titanium oxide for paints, where it has a greater whitening effect than lithopone or white lead carbonate, because of its permanence and its ability to reflect white light.

22

SOME MINOR METALS
and a few other elements used in metallurgy

ANTIMONY

In appearance this semimetal is similar to zinc. Being hard and brittle (it can be powdered with a hammer), it is rarely used alone, but is employed as a hardening addition to solder, pewter, white metal bearings, accumulator plates and telephone sheathing. Considerable quantities of antimony trioxide are used in flame-retardant compositions.

ARSENIC

Like antimony, this element is not a true metal, though it has some metallic properties. Throughout the long history of metallurgy, arsenic ores have been found in conjunction with those of copper, so some arsenic was contained in the copper when it was smelted. It was found to be advantageous, in conferring increased hardness when the copper was cold worked. Arsenic, as well as boron and phosphorus, is used as a dopant in the manufacture of silicon integrated circuits.

BARIUM

This silvery-white metal is spontaneously flammable in moist air and, though valuable in the form of some of its compounds, has only a few direct uses in metallurgy. Compounds known as ferrites (not to be confused with ferrite, discussed on page 132) are manufactured for use as magnetic components of electrical equipment. They contain the iron oxide Fe_2O_3 combined with the oxide of barium, or other oxides such as those of strontium or lead.

BERYLLIUM

This metal is lighter than aluminium and nearly as light as magnesium; it has a melting point of 1279°C, good corrosion resistance at temperatures up to 500°C and a high strength-to-weight ratio. It was once considered to be a wonder-metal, with a promising future for the construction of aircraft fuselages and space vehicles. Unfortunately difficulties were encountered in processing the metal, and it has toxic properties. Therefore beryllium-rich alloys are rarely used, but there is a useful range of copper alloys containing beryllium. A copper alloy with 1.8 per cent beryllium which has been rolled, heated to 800°C and quenched in water has a strength as great as that of mild steel. If this quenched alloy is reheated to 335°C and then cooled, precipitation hardening (see page 159) takes place and the alloy develops a strength twice that of mild steel. Beryllium–copper is among the strongest non-ferrous alloys.

BISMUTH

Nearly half of bismuth's production is required as chemical compounds in cosmetics, pharmacy, flame retardants and as a catalyst in the manufacture of synthetic fibres, rubbers and plastics. About a third is used as an addition to malleable cast iron to improve its strength. Another important use is in the low melting point alloys, discussed on page 41.

BORON

Principally this non-metal is used in the form of borax (sodium borate combined with molecules of water) and boric acid, required for the manufacture of porcelain, enamels and glass in dying textiles, fireproofing timber, as a food preservative and for pharmaceutical purposes.

Boron has a great capacity to absorb neutrons, so steel containing boron has been used in nuclear power stations for control rods (see page 257).

In ferrous metallurgy, small quantities of ferro-boron are added to improve 'hardenability' (the degree to which a steel will harden when quenched). Great hardness is a quality associated with the metallurgical uses of boron, and relatively cheap steels may be 'boronized' to make surfaces more hard-wearing than those produced by carburizing or nitriding.

Compounds formed between boron and titanium or zirconium are practically as hard as diamond; they are better conductors of electricity than some pure metals and have very high melting points, of around 3000°C. They are resistant to corrosion by molten metals and salts, and, now that they are becoming commercially available, are expected to find many new applications.

Boron is an extraordinary element. Although it is a non-metal its tensile strength is greater than that of any pure metal. In addition to the uses mentioned above, it is one of the elements present in the latest high-power magnetic materials (page 270). It features in the process of rapid solidification (page 273) and in the superalloys (page 192).

CADMIUM

As cadmium minerals do not occur in sufficient quantities to justify mining them in their own right, the metal is obtained as a by-product of zinc production, with a yield of about 3 kg of cadmium for every tonne of zinc. Cadmium coatings are applied to steel, brass, copper and aluminium to protect them from corrosion, especially for components which are exposed to marine conditions. Nickel–cadmium batteries are manufactured in a great variety of types and sizes for emergency power in hospitals, telephone exchanges, computer installations – in fact, wherever an uninterrupted power supply is essential. Small nickel–cadmium batteries have many applications, in portable telephones, camcorders, power tools and domestic cordless appliances.

Pigments based on cadmium sulphide and cadmium sulphoselenide can be produced in a range of brilliant colours, resistant to chemical attack and to degradation by light. The millions of photoelectric cells used to provide 'turn on' and 'turn off' times for street lighting are activated by cadmium sulphide batteries. Cadmium-bearing stabilizers consist of a mixture of cadmium and barium organic salts, incorporated in polyvinyl chloride (PVC) to ensure a good initial colour and long service life.

CAESIUM

This metal possesses the highest thermal expansion – about eight times that of iron. Apart from mercury, it has the lowest melting point of all the metals, 28.6°C. It is used in extremely sensitive light-detectors, and in atomic clocks, which measure time intervals much less than a millionth of a second.

CALCIUM

Some of calcium's chemical compounds are well known, for example limestone and chalk are calcium carbonate, though the 'chalk' for writing on blackboards is calcium sulphate. A lead alloy containing 0.09 per cent calcium with 0.3–0.5 per cent tin features in modern maintenance-free batteries, discussed on page 183. Calcium is a soft metal and corrodes rapidly in air, so it is not used on its own for engineering purposes, but its metallurgical importance has increased since it was brought into use in the processes for winning chromium, thorium, uranium, zirconium and some of the rare earth metals from their oxide ores.

CHROMIUM

The electroplated layer of chromium about 0.3 micrometres thick that is deposited over nickel plate on steel, brass and zinc alloy is familiar to everyone. By far the most important use, however, representing about 70 per cent of the metal's production, is for high-performance steels. Additions of from 11 to 30 per cent chromium give the range of stainless steels, discussed on page 141. The chromium is provided as ferrochromium containing about 70 per cent of the metal, 4 to 6 per cent carbon and the rest iron. This has a lower melting point than the pure metal, and is a cheaper and more convenient way of adding it.

There are several other alloy steels containing chromium, together with other elements such as molybdenum and vanadium, which can be used at temperatures of up to 700°C for components of, for example, steam turbines. Resistance wires for electrical heating elements contain 80 per cent nickel and 20 per cent chromium. This composition was the base from which the range of Nimonic alloys was developed for jet engines.

COBALT

This metal has a wide range of applications, but most of them involve cobalt as an alloying element, usually in small percentages in superalloys and Nimonics (pages 192 and 191 respectively). There is a wide range of magnetic alloys: until recently the alloy with 35 per cent cobalt, 57 per cent iron, 5 per cent tungsten, 2 per cent chromium and about 1 per cent carbon was the most highly magnetic material known. Since then there have been outstanding developments, thanks to the association of cobalt with various of the rare earth elements discussed on page 215. Among these, a cobalt–samarium alloy has been at the head of the list, though alloys containing neodymium have entered the field and there remain many combinations still to be explored.

A group of cobalt-rich alloys known as Stellites are extremely hard and resistant to corrosion, and retain these properties at high temperatures. A typical alloy contains 68 per cent cobalt, 26 per cent chromium, 5 per cent tungsten and 1 per cent carbon. Such alloys are required for petrol and diesel engine exhaust valves, fluid flow control valves and extrusion dies. A minor, but to many sufferers important, use of Stellite is in the rebuilding of knee and hip joints.

Cobalt therefore plays an important part in several modern technologies. Over half the world's supplies come from Zaire and Zambia, which have had their full share of troubles, so the price and availability of cobalt have fluctuated dramatically. More recently supplies have become available from the former USSR

GALLIUM

A soft, silvery-white metal, gallium is unusual in having a wide range between its melting point, 29.78°C, and its boiling point, about 2070°C. Gallium is one of the elements that have been used for doping silicon for transistors. The development of miniature gallium arsenide lasers led to a revolution in telecommunication technology, and these devices also feature in compact disc players. The compound gallium phosphide is used in light-emitting diodes, tiny electro-luminescent 'lamps' which operate in clusters to form digital read-out displays.

GERMANIUM

This semiconductor element was one of the first materials used for transistors, which are amplifiers of small size and great reliability using only a millionth of the power required by the thermionic valves which they replaced. Transistors were developed as a result of research in the 1940s, in which semiconductor materials were doped with minute amounts of other elements, producing an enormous change in their electrical properties. Electrodes were attached to a single crystal of germanium, which was doped with arsenic at one electrode and with gallium or indium at another. This diode has different electrical conductivities in opposite directions. By fixing two diodes together, back to back, a transistor is formed. As will be seen when we look at silicon, that material has replaced germanium for most transistors.

However, germanium is sometimes preferred for specialized applications such as infrared and gamma-ray detectors.

GOLD

To produce one troy ounce (about 32 g) of gold, 3 tonnes of ore must be mined from the reefs, conveyed to the shaft, hoisted to the surface and crushed to the consistency of face powder. The ore is put into large tanks containing a dilute solution of potassium cyanide. It is agitated with compressed air, and the gold dissolves in the cyanide solution. Then the solution is filtered, and the gold is precipitated from it. In addition to the 3 tonnes of ore, about 5000 litres of water, 750 kWh of electric power, explosives, potassium cyanide, compressed air and 39 man-hours are required to produce the one troy ounce of gold. There are some comments about deep mining on page 13.

Gold has complete resistance to atmospheric corrosion, as was shown when the tomb of Queen Pu-Abi was excavated in Iraq 4600 years after her death. When Tutankhamun's tomb was discovered by Howard Carter after 3300 years, the gold and jewelled coffin, the face mask and the exquisite jewellery were untarnished.

Gold is so ductile that one troy ounce can be drawn into a wire 40 kilometres long. It is so malleable that it can be beaten into leaf so thin that it is transparent. To produce gold leaf, the metal is first rolled to a thickness of 0.025 mm, and cut into squares which are then interleaved with layers of a skin-like material and arranged in stacks of about 400, which are then beaten down to reduce the thickness. The process is repeated over and over again until the leaf is only 0.2 micrometres thick – about one-thousandth the thickness of this page – and so delicate that a special 'gilding tip' brush must be used to lift and apply it. Gold leaf can be admired in many places from the name of the local pub to the flame of the Statue of Liberty and the magnificent rooms of the Palace of Versailles. The vital electronic systems that enabled man to take his first step on the Moon were protected from take-off blast by coatings of gold, and it was used in the circuitry of computers on the Moon voyage.

For many purposes the pure metal would be too soft on its own, so it is hardened by the addition of alloying elements, copper being the most common; silver, nickel and palladium are also added for making jewellery. Copper alone produces reddish alloys, while with silver alone the colour tends towards greenish-white. Together they produce the familiar rich yellow colour. In Britain the gold content of jewellery alloys is designated by the carat system. Pure gold is 24 carat; 18 carat contains 18 parts of gold and 6 parts of alloying elements. The compositions used in the UK are 22, 18, 14 and 9 carat. Gold offered for sale must be hallmarked to one of these legal standards. Hallmarking is the oldest of all consumer protection systems, having been carried out by the Worshipful Company of Goldsmiths in the City of London since the year 1300. In Europe and elsewhere the gold content is expressed decimally; thus 9 carat gold is known as 375.

Rolled gold consists of a thin layer of gold alloy bonded on a base metal such as brass or nickel–silver. The process involves high temperature and pressure, and the final product has a perfect adhesion between the two metals, matching the appearance of gold but without its high cost. Recently, rolled gold has tended to give way to sprayed gold, which also looks like 'the real thing' but is less expensive.

HAFNIUM

Zirconium ores contain hafnium compounds in minute quantities; the two elements are chemically similar, and at first it was difficult to separate them, but the development of the solvent extraction process (page 16) made this easier to do. Hafnium is mostly used as an alloying element, for example with molybdenum, niobium and tantalum. It is very effective for absorbing neutrons and it has good resistance to corrosion in hot water. Hafnium clad in stainless steel is used for control rods in some nuclear power stations.

INDIUM

Although it is widely distributed in the Earth's crust, indium does not occur anywhere in concentrations that would allow the metal to be produced in its own right, so it is obtained as a by-product in the smelting of zinc, copper, lead and tin. In 1928 the world production was only one gram, but this has now climbed to about 150 tonnes a year. It is a versatile metal with applications in electronics, where low melting point solders based on indium are invaluable. Fusible alloys containing indium are used in medical radiology and safety devices. Intermetallic compounds of indium with phosphorus, antimony and arsenic generate electric current in the presence of light or, conversely, emit infrared light when current flows across wafers of the compounds. Ultra-high purity grades of the metal are being developed for semiconductors and optical fibres.

Indium is used for the surface coatings of some bearings, to give improved resistance to corrosion and wear. A three-layer bearing is produced: the back, of low carbon steel, is lined with a copper–lead–tin alloy which, after broaching or boring, is plated with a layer of pure lead and machined accurately to size; a thin layer of indium is then deposited on the lead. The bearing is heated and the indium diffuses through the bearing, almost to the interface between the steel back and the lining. The function of the indium here is threefold. First, it reduces the susceptibility of the copper–lead layer to the corrosive attack of lubricating oils; second, it gives increased fatigue strength to the lead overlay; and third, it provides good anti-friction properties. The use of indium-faced bearings has extended to medium-heavy diesel engines, and in the coming years bearings larger than 200 mm, the present manufacturing limit, will be achieved.

In the search for improved anti-ice and anti-mist materials for laminated aircraft windscreens, Triplex have treated an alloy of indium with tin by sputtering. The composite oxide so formed is deposited within the laminate to a thickness of about 2.5 micrometres, about a hundred times thicker than the gold layer previously used, but this deposit is transparent and of a more neutral colour than the gold film.

IRIDIUM

With its specific gravity of 22.56, iridium, a member of the platinum group of metals, is second only to osmium in density. Iridium's corrosion resistance is even better than that of platinum, but despite a very high melting point of 2454°C, it is less oxidation resistant than either platinum or rhodium. Iridium is difficult to work, and only recently has it begun to be used in its elemental form. Iridium-tipped spark plugs have been developed to replace platinum-tipped plugs in aero engines. Platinum alloys containing up to 30 per cent iridium are used for jewellery; the 10 per cent alloy is used for diamond settings in the USA. Platinum–iridium alloys are used for standards of weight and length because of their great stability and permanence.

After exploring the atmospheres and moons of Jupiter and Saturn, the space probe Voyager 1 left the solar system. Its sister, Voyager 2, reached Uranus in 1986, and Neptune in 1989. Both craft continue to survey the solar system from the outside. The Voyagers were provided with heat sources consisting of three sets of eight ceramic spheres of plutonium dioxide encapsulated in thin iridium shells, with vents to release the gas products of radioactive decay.

LITHIUM

This metal has only half the density of water. It reacts vigorously with air and water but, in spite of such problems, the new aluminium–lithium alloys discussed on page 270 are being made successfully and used in aeronautics. The low density and atomic weight of lithium, combined with its high reactivity, make it an attractive material for the electrodes of batteries that are required to provide a lot of power for their weight. Various formidable practical problems have been overcome, and the output of lithium batteries has continued to expand, particularly in Japan.

Two isotopes (page 250) of lithium are significant in nuclear engineering. Lithium-7 does not absorb neutrons in nuclear reactors, but lithium-6 has a great ability to absorb neutrons, and hydrogen isotopes are derived from it. Perhaps in the twenty-first century the world's energy problems will be solved by the 'taming' of thermonuclear fusion, for which experimental equipment is now being built. Lithium's properties of low melting point and high boiling point referred to above have already made the metal suitable as a heat transfer agent in some nuclear reactors, and it may be chosen to extract and transfer the colossal heat from thermonuclear reactors. Furthermore, the fusion reactors of the future may rely on lithium which would be broken down into helium and tritium.*

MANGANESE

Each year about 25 million tonnes of manganese ore is mined, containing on average 35 per cent metal. It is therefore fifth in tonnage, after iron, aluminium, copper and zinc. The metal is not used on its own but is extracted in the form of ferro-alloys, including one with 80 per cent manganese and another with 20 per cent manganese and 10 per cent silicon. Such alloys are added to steel in the course of manufacture; in fact, more than 90 per cent of the metal produced is destined for the steel industry.

One important feature of manganese is that it reduces the deleterious effect of sulphur in steel. Iron sulphide forms continuous films which cause the metal to become embrittled. When manganese is present it combines with the sulphur to form manganese–iron sulphide, which is

---

* The tritium atom is a hydrogen atom with two neutrons as well as a proton in the nucleus.

distributed in small particles and is not harmful. Most low alloy steels contain 0.2–1.2 per cent manganese to provide toughness. With 1.3–1.8 per cent the strength of steels is further increased; such materials are used in automobiles, where the greater strength permits weight reduction. Steel for bridges and ship plates contains 1.2–1.6 per cent manganese with 0.15–0.20 per cent carbon.

Very hard steels with 10–14 per cent manganese have been mentioned on page 140. They are used, in the water quenched condition, where great resistance to abrasion is necessary: in railway points and crossings, rock drills and stone crushers.

The so-called manganese bronze does not necessarily contain manganese. The Victorians knew a thing or two about publicity and they found that when a high grade brass containing 40 per cent zinc was called 'manganese bronze', it sold very well.

The metal is a useful alloying element with aluminium, nickel, silver, titanium and copper-based electrical resistance alloys.

Some copper–manganese alloys have the property of damping out vibrations; one alloy with 53 per cent manganese, 38 per cent copper and small amounts of aluminium, iron and nickel is suitable for casting and is used for the manufacture of propellers for submarines.

MERCURY

The only metal that is liquid at ordinary temperatures is mercury, sometimes known as 'quicksilver'. It is nearly twice as heavy as iron; it has uniform volume expansion over the entire temperature range of its liquid state, which partly accounts for its use in thermometers. It is toxic, and consequently there has been a trend in recent years to reduce the amount of mercury in use. However, there are many important applications where replacement is difficult. Its principal use is in the manufacture of alkaline dry-cell batteries with a mercury oxide cathode, which have long life and function well in hearing aids, calculators, watches and cameras. In the USA about 600 000 kg of mercury is required for such batteries, but many manufacturers are gradually decreasing the amount of mercury in these products. Mercury dissolves other metals without being heated, forming amalgams. In dentistry, a silver–mercury amalgam has been regularly used; the profession is convinced that it is not harmful, but experiments are being carried out, so far without great success, in the use of plastic materials.

MOLYBDENUM
(*commonly called 'Moly'*)

The largest ore deposits and metal production operations are in the western hemisphere, with the USA contributing the biggest share, followed by Canada and Chile. At Climax, Colorado, 3500 metres above sea level, more than 500 tonnes of ore is mined to recover each tonne of molybdenum metal.

Molybdenum has a high melting point, 2625°C, which is exceeded only by tungsten, rhenium, tantalum and osmium. It has similar properties to tungsten, but is more easily workable, and lighter. Molybdenum's high corrosion resistance leads to its use in the chemical industry for such parts as autoclaves and tubes, while in the glass industry it is used for electrodes and tubes for the transportation of molten glass.

The steel industry absorbs about 75 per cent of all the molybdenum produced, mostly for low alloy steels for construction uses. When added to alloy steels containing nickel and chromium, molybdenum reduces the tendency to embrittlement that would otherwise occur when such steels are heat treated. Among many other new developments, molybdenum has played a part in improving the efficiency of the rail traffic hauling iron ore. In one Australian mine there are ten trains, each consisting of about 140 cars, each of which can carry about 100 tonnes of ore, travelling at speeds up to 50 kph. Ordinary carbon steel rails were wearing out rapidly under such massive and continuous loads; chromium–molybdenum alloy steel rails have replaced the carbon steel ones, with significantly better results.

Over a fifth of the molybdenum required by the ferrous industry goes as small additions, up to 3 per cent, to stainless steels. In another related use, the pipelines that convey oil and gas from Alaska, northern Canada and Siberia are made from nickel–chromium–molybdenum steels which can withstand temperatures as low as −60°C.

Research work in the USA led to molybdenum replacing tungsten in high speed steels; it is generally reckoned that molybdenum has twice the 'power' of tungsten, so that a steel previously containing 18 per cent tungsten could be replaced by one with 9 per cent molybdenum. In practice, chromium and vanadium are also added to high speed steels. The inclusion of the metal in HSLA (high strength, low alloy) and dual phase steels has already been mentioned on page 144.

There are numerous small but important uses of the metal, such as for radiation shielding in vacuum melting furnaces and as a construction material in nuclear engineering. A compound of molybdenum has important uses in the field of lubrication: molybdenum disulphide is an effective addition to industrial and automobile lubricants which have to withstand high pressures and temperatures.

NIOBIUM
(known as columbium in the USA)

Ores of this metal are always found in association with those of tantalum. Indeed, it is so named because Niobe was the daughter of the cruelly punished King Tantalus. Most of the world's production comes from Brazil, but Zaire and Canada possess small deposits. Niobium has a melting point of 2468°C, so is suitable for components of high temperature furnaces. When alloyed with 1 per cent zirconium, it has twice the strength of the pure metal at 1000°C, and small tubes of this alloy are used as electrode supports for high pressure sodium lamps in street and motorway lighting. Niobium does not absorb neutrons readily, thus making it suitable for support brackets and fuel cans in nuclear reactors.

The uses of the pure metal mentioned above represent only about 5 per cent of the total production. The remainder is extracted as what is universally called ferro-columbium, required for alloy steelmaking, especially for the HSLA (high strength, low alloy) steels used for gas and oil pipelines and in automotive engineering. Niobium oxide, Nb_2O_5, is required in the manufacture of several types of glass.

At very low temperatures, typically below $-263°C$, some niobium alloys and compounds become superconductive – they lose all resistance to electric current. These materials include a niobium alloy with 42 per cent titanium and a niobium–tin intermetallic compound, Nb_3Sn. During the last few years dramatic discoveries have been made in the field of superconductivity, making the niobium materials of historic but not commercial interest. These developments will be discussed on page 293.

OSMIUM

The main interest of this rare metal lies in its neck-and-neck contest with iridium to be the heaviest of all elements. When a small porosity-free piece of osmium made by arc melting in a vacuum is tested, its specific gravity is found to be 22.59. A piece of iridium sheet has been measured to have a specific gravity as high as 22.56. On this count osmium wins. However, from theoretical calculations of the lattice spacings of the two metals, osmium should have a specific gravity of 22.60 and iridium 22.65. Until a piece of iridium has actually been produced that is heavier than osmium, we, and no doubt the *Guinness Book of Records*, will continue to award the trophy to osmium. In its elemental form osmium is virtually useless, as it forms an oxide which volatilizes at room temperature. The metal got its name from the Greek word for 'a smell' because its oxide possesses a peculiar odour.

PALLADIUM

A member of the platinum family of metals, palladium is the one most similar to platinum, though it is slightly less corrosion resistant and has a lower melting point. Palladium is cheaper than platinum, has a lower density and can replace it in some applications where conditions are not extreme, including electrical contacts in speech circuits of telephone systems. Palladium catalysts are used in oxidation and hydrogenation processes, in margarine manufacture and the production of ethylene from acetylene. This metal has the unique property that it can rapidly absorb up to 900 times its own volume of hydrogen. Super-pure hydrogen for chemical and metallurgical processes is obtained by passing impure hydrogen through a diaphragm of heated palladium which acts as a filter, excluding other gases.

We cannot resist mentioning that a few years ago some research physicists in Utah claimed that they had achieved 'cold fusion' with the aid of this extraordinary property of palladium. Great excitement ensued in the media because such an invention might provide energy safely and cheaply. The idea was this. A plate of palladium is coated on one side with palladium oxide. It is then placed in a sealed container from which the air has been extracted. Deuterium* is pumped into the container and the palladium absorbs a vast amount of the gas. The plate is removed, coated with gold and reheated, whereupon some of the deuterium atoms 'fuse', producing an isotope of helium and evolving energy, which heats the plate by up to 100°C. Unfortunately when the experiments had been repeated in many parts of the world, it appeared that the claims were unjustified. Now it is almost certain that this method will not, alas, solve all our energy problems.

---

* The deuterium atom is a hydrogen atom with a neutron as well as a proton in its nucleus.

PLATINUM

Today the principal supplies outside Siberia are from South Africa, where it is recovered from a sulphide ore, and Canada, where it is associated with the Sudbury nickel/copper ores. Platinum has exceptional resistance to corrosion, and has a high melting point, 1773°C. It is therefore required where even the slightest amount of oxidation or corrosion would be detrimental; for example, for electrical contacts carrying small voltages, crucibles and other pieces of apparatus used in critical chemical analyses. Platinum alloyed with copper or iridium has become popular as a jewellery alloy, particularly for the mounting of diamonds on rings. The metal is hallmarked with an orb.

Platinum catalysts are used in many chemical processes, such as catalytic oxidation to form ammonia, with large woven gauzes of platinum alloyed with rhodium. A few years ago pure platinum was used as a catalyst in reforming low-octane petrol, but now the metal is alloyed either with small amounts of rhenium or with silicon and germanium.

Efforts to reduce pollution and smog caused by hydrocarbons, carbon monoxide and oxides of nitrogen have involved the use of catalysts containing platinum, palladium and sometimes rhodium in automobile feedback systems, which self-adjust according to the composition of the exhaust gases: these are the so-called 'catalytic converters'. The fuel must be lead-free, because leaded 'gas' poisons the effect of the catalysts. Each manufacturer has an individual approach to the subject, with the hope that something less expensive than these costly metals will prove effective. In the meantime, the automobile industry has become the largest user of platinum; in a typical converter about two grams of each of platinum and palladium are required.

To start the reaction, the catalysts must first warm up to about 300°C, so they are usually mounted close to the engine. The carbon monoxide is converted to dioxide, while hydrocarbons are converted to water and carbon dioxide. In a 'three-way' converter, the exhaust gases come into contact with rhodium, which reduces the noxious nitrogen oxides to nitrogen and oxygen.

Optical glass is melted in large platinum-lined crucibles, so as to avoid contamination by refractories; platinum-clad apparatus is used in many other glass-making processes. Continuous filament glass fibre is made by melting the glass in platinum–rhodium alloy vessels, which are heated by electrical resistance, the filaments being drawn off through nipples in the bottom of the vessel. Pyrometers, instruments capable of measuring high temperatures up to about 1400°C, are of platinum/platinum–rhodium alloy.

PLUTONIUM

Plutonium was first made synthetically in March 1941 by bombarding uranium atoms in the cyclotron at the University of California. The bombardment led to the formation of neptunium, which is unstable, and thence to plutonium. During the operation of nuclear reactors plutonium is formed from uranium; it can be separated chemically, and is then available for use in fast-breeder reactors. Plutonium is very heavy – twice as dense as lead. It looks like nickel, but tarnishes in a few minutes and quickly turns into a powdery green plutonium oxide. If the air is damp it may catch fire spontaneously – and the smoke is deadly. Plutonium is highly reactive, fiendishly toxic and a bone-destroyer.

POTASSIUM

This reactive, soft, silvery-white metal has chemical properties similar to those of lithium and sodium. At about fifty–fifty composition potassium and sodium form a eutectic melting at room temperature, and this alloy is used for cooling in consumable-arc titanium melting furnaces, like the one shown in *Plate 25*. Potassium compounds are widely used in chemical processes, including the extraction of gold by potassium cyanide.

RADIUM

This extremely rare, highly radioactive, heavy metal is extracted from uranium ores, where it occurs in the proportion of about one part radium to three million parts uranium. The fantastic scarcity of radium may be assessed from the annual production in Canada, the world's largest source of supply, which amounts to only a few kilograms. The metal is chiefly used for medical purposes on account of the penetrating radiation that it emits continuously, which is used for the treatment of cancer and some skin diseases. The historical interest, the medical value of radium and the inspiration of the story of Marie Curie's endeavours are very great – but metallurgically, radium is of little significance.

RARE EARTH METALS

From time to time mention is made in technical journals of elements with tongue-twisting names, like praseodymium, which are arousing new interest. They represent about one-seventh of all the known elements but are scarce, expensive and produced in only small quantities. They are known also as lanthanides, because lanthanum, with an atomic weight of about 139, is the first of the series; lutetium, with an atomic weight of about 175, is the last. They are all somewhat similar to aluminium in their properties. In alphabetical order, they are:

cerium	gadolinium	neodymium	terbium
dysprosium	holmium	praseodymium	thulium
erbium	lanthanum	promethium	ytterbium
europium	lutetium	samarium	

Scandinavian scientists played an important part in isolating these metals, and the names given to several of them reflect this influence. The mineral gadolinite was named after the Finnish chemist Johann Gadolin, hence the name gadolinium. The mineral had been discovered near the Swedish town of Ytterby, from which the names ytterbium, erbium and terbium were derived. Holmium was named from Stockholm, and thulium from Thule, the early name for far-northern Europe.

The Russian mining officer Samarski lent his name to posterity with the metal samarium. Lutetium was named from Lutetia, the ancient name for Paris, and europium derived its name from Europe. Cerium was named after the asteroid Ceres.

Some of these metals borrowed Greek roots for their names. Dysprosium means 'the one hard to get at'; lanthanum 'the hidden one'. The name didymium, meaning 'twin', was given to a material which was found to be composed of two elements which, when isolated, were called neodymium, 'the new twin', and praseodymium, 'the leek-green twin'.

The term 'rare earth' may be misleading. There is more cerium than tin in the Earth's crust, and all these elements, apart from promethium, are more abundant than platinum. The problem is that the percentage of metal contained in the rare earth minerals is so small that their extraction is tedious and costly.

One of the earliest uses of a rare earth metal was in cigarette lighter flints. The so-called 'misch metal' is an alloy containing cerium and other rare earth metals. When the alloy is rubbed by the steel wheel of the lighter a substantial spark is formed; the alloy is said to be 'pyrophoric'. Cerium oxide or mixed rare earth oxides are used as abrasives for high quality polishing of the glass of lenses, mirrors and TV face plates. Small additions of cerium make it possible to produce spheroidal graphite cast iron (page 148).

Several rare earth metals are used as catalysts in the petroleum industry. Europium activates an yttrium phosphor, emitting an intensely red light.

More and more of these metals are entering into high-tech industries: gadolinium for control rods in some nuclear reactors, and samarium and neodymium in the new magnetic alloys (page 270).

The family of rare earth metals has been termed a Pandora's box, but it has been pointed out that, so far, the list includes neither 'delirium' nor 'pandemonium'.

RHENIUM

This very heavy, silvery-white metal has the second-highest melting point, 3180°C, and a resistance to corrosion similar to that of the platinum group of metals. Rhenium strengthens tungsten and molybdenum at high temperatures, and improves their ductility, thus easing the production of high temperature thermocouple wires in tungsten–rhenium and molybdenum–rhenium alloys.

RHODIUM

When William Hyde Wollaston succeeded in separating a previously unknown metal from platinum in 1804, he decided to call it rhodium because some of its compounds were coloured rose-red; the Greek word for rose is *rhodon*. The metal is one of the most expensive. About half is mined in the countries of the former USSR, and nearly half in South Africa; only a few grams of rhodium can be extracted from each tonne of the ore.

Rhodium has similar properties to those of platinum, including great resistance to corrosion. Platinum–rhodium alloys are used as catalysts in the production of nitric acid from ammonia. Thermocouples for measuring high temperatures up to 1800°C consist of platinum with platinum–rhodium alloy. For still higher temperatures, up to 2100°C, thermocouple wires are of iridium with iridium–rhodium alloy. Rhodium, associated with platinum and palladium, is used in automobile catalytic converters to decompose the noxious nitrogen dioxide. The metal is electroplated to give an enduring surface on white gold and silver jewellery and on heavy duty electrical contacts.

RUTHENIUM

This metal is brittle so is not used on its own, but it is sometimes alloyed with platinum and palladium. Ruthenium electrodeposition processes have been developed in recent years and, although the dark colour precludes the use of these deposits for decorative purposes, their hardness and corrosion resistance approach that of rhodium.

SCANDIUM

The principal use of this rare element is in the manufacture of 'metal halide' lamps based on scandium iodide. They give an efficient light output with a colour close to that of sunlight, so when used for television presentations from the studio or an outside broadcast at night, the colour gives the scene the appearance of daylight.

SELENIUM

When light falls on selenium, its electrical conductivity increases by many orders of magnitude, depending on the intensity and wavelength of the light. Selenium is not unique in this respect; all semiconductors exhibit it to some extent. Advantage was taken of the property in the operation of early Xerox copier machines, but now they use organic photoconductors. Temperature measurement devices which operate by registering the amount of light emitted by a hot material rely on selenium. About 40 per cent of the world's demand is represented by the use of selenium for decolorizing the green tint caused by iron impurities in glass bottles.

SILICON

Although not a metal, silicon was bound to be mentioned many times in this book because of its important roles in ferrous and non-ferrous metallurgy and its ever-increasing use in computers and other electronic equipment. About 1 per cent of silicon added to manganese confers improved elasticity on steels used for bridges and car springs. Silicon is a principal alloying element in steels used for electric motor laminations, transformers and generators. The important aluminium alloys containing silicon are discussed on page 155.

Silicon is manufactured in powerful electric arc furnaces. The raw material is silicon dioxide, in the form of quartz. The reducing agents are carbonaceous materials – coke and wood chippings. As in most smelting operations, the reactions are many and complex, but the basic chemical equation is $SiO_2 + 2C \rightarrow Si + 2CO$. The end product is silicon of about 85 per cent purity. This is crushed to a powder and then leached, to arrive at a purity of about 98 per cent, suitable for many uses in metallurgy. Another route is the smelting of mixed iron and silicon compounds to make ferro-silicon. Ultra-high purity silicon, required for silicon chips, is made by first reducing silicon compounds with hydrogen, followed by zone refining (described on page 300), which reduces the level of impurities to less than one part in a hundred million.

To make integrated circuits, single crystals of silicon are cut into thin wafers and the surface of each wafer is coated with a photoresistant polymer, so that when light falls on a predetermined area, its chemical composition is changed and it can be selectively dissolved away. The surface of the wafer is then chemically treated through the pattern of exposed holes. The illumination pattern is a mask made by depositing a thin chromium film on glass, so that an image can be formed on the surface of the silicon by an instrument like a slide projector. Several different masks are needed to construct the integrated circuit.

The silicon chips are made by sequentially treating the surface in various ways, using the lithography process to etch the required patterns. These treatments include oxidizing the silicon surface to provide an insulating layer of silicon dioxide, and doping the silicon with phosphorus or boron to give the required electrical conductivity. The oxidation is performed at about 1000°C; doping is achieved by bombarding the surface with a beam of high energy ions. Then metals such as aluminium are evaporated onto the surface in a high vacuum to provide the electrical contacts. There are many stages in the production of a silicon chip from a single crystal of the element, each stage of which requires complex equipment and a meticulously controlled environment.*

For thirty years the accuracy of the photolithography process has steadily increased, so that patterns can be formed on the surface of the silicon wafer with an accuracy of less than 1 micrometre. Because the individual devices are so small, a wafer 20 centimetres in diameter can contain up to a thousand million circuit elements. Each wafer is cut into about a hundred separate integrated circuits. Personal computers have their entire central processing unit made from a single integrated circuit containing from 100 000 to a million transistors. These computers also need memory chips, whose capacity has increased tremendously in the past decade. The most widely used memory chips at present have a capacity of 4 megabytes – sufficient to store information equivalent to about ten copies of this book. By the end of the century the capacity is expected to increase to 256, or even 512 megabytes.

---

* There is a demonstration of the making of a silicon chip at the Tech Museum of Innovation at San José, in the heart of California's Silicon Valley. This splendid museum has many other fascinating exhibits, including a demonstration of a shape-memory alloy (page 271).

SILVER

Although a considerable amount of silver is produced from the sulphide ore argentite, nearly half the world's requirements are supplied as a by-product from the extraction of other metals, and as metal reclaimed from demonetized silver coinage, industrial waste and the photographic industry, which is the largest user of silver. Most photographs start with an emulsion that uses a light-sensitive silver bromide, chloride or iodide, or a combination of the three. Developing changes the exposed halide to metallic silver, which forms the final image.

Although few silver coins remain in circulation, a historic use is noteworthy. Maundy money, distributed in England by the Queen on the Thursday before Easter, contains 92.5 per cent silver. Special coins of 1, 2, 3 and 4 pence in value are minted for the occasion. The amount given is determined by the monarch's age, one 'penny' for each year, and the number of recipients is governed by the same factor, one woman and one man for each year.

Silver has the highest reflectivity of light and the highest electrical conductivity of all metals, though its tendency to tarnish and its comparatively high cost limit its use in these fields. The silver used for jewellery, cutlery, teapots and other decorative domestic articles is subject to hallmarking. In the UK there are two legal alloys: Britannia silver contains 95.84 per cent silver, and the much more common Sterling silver contains 92.5 per cent; the alloying element is usually copper.

A recent development is Hardenable silver, which allows the silversmith to produce decorative articles using conventional working techniques and then, by a subsequent heat treatment, to increase their strength. Hardenable silver contains over 99 per cent of the metal, with small additions of magnesium and nickel. When heat treated in air or oxygen, the magnesium in the alloy oxidizes to produce minute particles of magnesium oxide which make the metal hard and rigid. All forming operations are carried out before heat treatment, while soldering or enamelling are done after.

SODIUM

This is a very soft, wax-like metal which reacts vigorously with water and corrodes rapidly in air. Strangely enough, when it is combined with another noxious element, chlorine, common salt is formed. Sodium melts at 97°C and, as liquid, has proved to be a superb material for conducting heat in the reactors of fast-breeder nuclear power stations.

Low pressure sodium streetlights, in which an electric current vaporizes sodium and makes it glow, were introduced in the UK in 1936 to replace mercury and tungsten filament lamps. In the early 1960s these began to give way to high pressure sodium lamps, containing an amalgam of sodium with other metals. Under high pressure, sodium vapour emits a much broader spectrum, giving a more natural light than the yellow glow of low pressure sodium. It now accounts for over 90 per cent of the lighting used for motorways and arterial roads and in towns. At first the aggressive properties of the sodium vapour caused the lamp enclosures to deteriorate, but then it was found that polycrystalline aluminium oxide, which had been used for coating the nose cones of rockets, proved effective in sealing and protecting the lighting electrodes.

The use of small amounts of sodium to modify aluminium–silicon alloys has been described on page 155, and is illustrated in *Plates 31c* and *d*.

STRONTIUM

A reactive metal, resembling barium and calcium, strontium is used in fireworks, burning with a crimson flame. It is also used in some magnetic ferrites (mentioned on page 201). The lethal radioactive isotope strontium-90 is present in the fall-out from nuclear explosions.

TANTALUM

The most important property of tantalum is its extremely high corrosion resistance, provided by a protective oxide layer; it has a high melting point, 2996°C. It has the highest capacitance (the ability to store electrical energy) of all the metals, and nearly half the tantalum made is used in powder or foil form as small capacitors in the electronics industry. By using tantalum it is possible to make capacitors one-tenth the size of normal aluminium-foil ones, and still maintain the same capacity. The market is enormous – Japan makes over four billion each year.

Tantalum is unaffected by the majority of acids, and is used to avoid corrosion problems in chemical engineering, where it is used for vessel liners, heat exchangers and thermocouple protection sheaths. It is valuable in surgery for bone splints, jawbone implants and other implants which must be left permanently in the body. The compound of tantalum and carbon is the most refractory metallic substance, having a melting point of over 4000°C. The tantalum carbide is added to tungsten and titanium carbide cutting tools to improve their performance.

TECHNETIUM

This radioactive element is never found in nature, but is entirely made by an atomic reaction. Its properties are similar to those of rhenium, but its melting point, 2200°C, and density are lower. Its most important use is in medical equipment for tracing centres of infection in the human body.

TELLURIUM

About half the world's small production of this semimetal is required for an alloying addition to steels that have to be machined, and to lead, where it improves resistance to vibration. About a quarter is used as a vulcanizing agent for rubber and as a component of catalysts for use in the manufacture of synthetic fibres. Cadmium–mercury–tellurium compounds are used in infrared detector systems.

THORIUM

This soft and ductile metal is closely related to uranium, and is one of the few naturally occurring elements that could possibly be utilized to develop atomic power by fission of one of its isotopes. Aside from future developments as a nuclear fuel, its principal present use is for adding to magnesium–zirconium alloys to improve the metal's properties at high temperatures.

TUNGSTEN

Robert Mushet, a friend of Henry Bessemer's, discovered that steels containing tungsten were capable of machining metals at high speeds, while retaining a good cutting edge. He made some of these 'self-hardening steels' with about 10 per cent tungsten, 1.2 per cent manganese and about 1.5 per cent carbon, but it was not till twenty years after his experiments that a satisfactory steel was introduced, thanks to the work of Taylor and White described on page 144. A typical modern high speed steel contains 18 per cent tungsten, 4.25 per cent chromium, 1.1 per cent vanadium and 0.75 per cent carbon. In other steels part of the tungsten is replaced by molybdenum; for example, one composition contains 6.4 per cent tungsten, 5 per cent molybdenum, 4.1 per cent chromium, 1.9 per cent vanadium and 0.82 per cent carbon.

When tungsten is combined with carbon, very great hardness is attained. Tungsten carbide is one of the hardest known substances and has a wide range of uses, including tool tips for high speed machining. It plays a vital part in tunnel drilling. *Plate 26* shows the scene in the vast 'crossover cavern', as big as Westminster Abbey, in the Channel Tunnel. The drilling machine, weighing 1350 tonnes and 265 metres long, had nearly a hundred tungsten carbide tips. It rotated at $4\frac{1}{2}$ rpm and advanced in cuts 1.5 metres deep, requiring the removal of over 80 tonnes of rock per cut.

For other situations there is a very wide range of tungsten carbide tool tips, each with a size and shape dictated by the needs of the cutting operation. *Plate 28a* shows one of the many cutting tools used for oil well drilling.

Although the tonnage represents only about 5 per cent of the total amount produced, the lighting industry requires tungsten for the billions of light bulbs used all over the world. The filaments are produced by powder metallurgy (Chapter 25); the tungsten powder is doped with small amounts of potassium–aluminium silicate and sintered at high temperature. The aluminium, silicon and oxygen from the silicate are evaporated, while the potassium remains in the tungsten, and this combination is drawn into wire.

Tungsten has the highest melting point of any pure metal (3380°C). It is extremely hard, and is about two and a half times as dense as iron. This combination of density and hardness has led to the popularity of tungsten darts. The expert player requires a dart within the 26 to 28 gram range, but those with brass bodies are stubby, making it difficult to get more than one in the same 'treble'. Elegant, slender dart bodies made by compressing tungsten powder with nickel (using processes described in Chapter 25) are only half as thick as the old-fashioned darts and are well suited to producing high scores, though they cost several times more than brass darts.

A group of alloys comprising tungsten with small amounts of nickel–iron or nickel–copper binders, known by the trade name of Densimet, have similarly high densities but are easier to machine. They are required for many applications where high inertial forces are encountered such as governors, gyroscope rotors and balancing weights. They are also suitable as shielding materials against X-rays and gamma-rays.

URANIUM

This metal, one of the heaviest elements (its specific gravity is 19.05, similar to that of tungsten) was discovered in 1798, and for over a century it was used as a colouring agent in glassware and ceramics. In the 1940s it was produced for weapons in which the amount of the fissile isotope uranium-235 had been increased. Today, virtually all the uranium mined is required as a fuel in nuclear reactors to generate electricity. Some can 'burn' uranium in its natural form, containing 0.7 per cent of uranium-235, but most reactors need uranium slightly enriched with 3–4 per cent of the 235 isotope. Nowadays the uranium is used in the form of its oxide.

Uranium minerals of economic importance, mainly pitchblende and carnotite, are widely distributed. Most ores are of low grade, with less than 0.2 per cent uranium, but there are deposits with up to 12 per cent metal content. The principal supplies in the 1990s are likely to be Canada, the USA, Australia, the countries of the former USSR, Niger and Namibia. Uranium ore concentrate is converted to uranium tetra-fluoride, and thence into uranium metal or the oxide for service in nuclear power plants.

VANADIUM

The existence of this element was deduced by a Swedish scientist, Nils Gabriel Sefström, in 1830, and it was named after Vanadis, the beautiful Scandinavian goddess of love. Forty years later the metal was isolated, but it was brittle and of little use; it was not until the twentieth century that a purer and more ductile form of vanadium was made. By the early 1950s the metal was being produced in sufficient quantity and quality to justify large-scale use in engineering. By far the greatest amount is required in the form of ferro-vanadium, made by treating vanadium oxide in the presence of iron in electric arc furnaces. The alloy is used for producing HSLA (high strength, low alloy) steels. These contain only small amounts of vanadium, between 0.05 and 0.1 per cent, some-times with even smaller amounts of niobium, but the strength of such steels is almost double that of 'straight' carbon steels.

Massive structures such as bridges, high rise buildings, spectator stands and aircraft hangars can be built economically, with the extra bonuses of reduced weight or longer spans, because of the strength of HSLA steel. Pipeline operators can make cost savings because it is possible to increase the pressure and thus to transport a greater amount of gas. To take one of many examples, the 1287 km pipeline from Prudhoe Bay in the north of Alaska to Valdez in the south is 1.22 m in diameter and has a wall thickness of 16.6 mm. Although the vanadium content was only 0.07 per cent, a total of 650 tonnes of the metal was used in the manufacture of the pipeline. An increasing amount of HSLA steel is required for submarine pipelines, especially to carry gas from North Sea platforms. About 5 per cent vanadium, in conjunction with chromium, is added to tool and die steels.

YTTRIUM

This rare metal is used in lasers in the form of a garnet, an yttrium–aluminium silicate. It can give very fast pulses – less than a picosecond (the prefix 'pico' indicating a millionth of a millionth). Such lasers may be used to make the kind of measurements required in the study of electronic circuits, to follow the extremely rapid motion of electrons. They also make it possible to monitor the changes that take place in chemical reactions.

Another recent application has been as a constituent of some nickel-based and cobalt-based superalloys; they have excellent resistance to corrosion and are used as coatings for turboblades in aero engines. Yttrium oxide, in combination with zirconium oxide, is used in the so-called lambda-sensors for determining the oxygen content of automobile gases.

ZIRCONIUM

Although this metal was isolated over 150 years ago it remained a laboratory curiosity until the difficult process was developed to separate it from hafnium, with which it always occurs naturally. In the form of finely divided powder or swarf, zirconium ignites at low temperatures and finds application in camera flashbulbs. Because of its corrosion resistance to alkalis, the metal is used in chemical engineering, for example as spinnerets in the rayon industry. Its importance as an alloying element with magnesium is mentioned on page 195. However, its largest metallurgical use is as Zircalloy, in the nuclear power industry (page 265). The use of zircon – the oxide of zirconium – as a high quality foundry sand is discussed on page 68. A large tonnage is required for refractories and ceramics.

23

CORROSION

Corrosion is the destructive attack upon metals by agents such as rain, sea water, polluted air or aggressive chemicals. Sometimes metals have to withstand severely corrosive conditions. Oil platforms such as those in the North Sea have been described as factories on legs; once in place, they have to remain in service for many years. Three-fifths of an oil platform structure is permanently under the water, reaching down below the sea bed; another fifth is alternately covered and uncovered by tides and lashed by high seas with waves several metres high. Above and below the surface of the water are webs of pipes, nuts, bolts, rivets and welded joints, all subject to the vile conditions of the environment. The lower parts of the structure become the home of shellfish, seaweed and other forms of marine life. As soon as the platforms are installed on the sea bed the waves begin to drive against them, putting tremendous pressure on the materials from which they are made. As a platform is battered in this way, a small area of corrosion which has developed under a mass of seaweed may develop into a crack, and the repeated impact of the sea may cause the crack to grow. In spite of these and many other problems, oil reaches the mainland in substantial quantities. This success results from the skills of engineers and metallurgists who have developed the right grades of steel and other metals, welding techniques, platform designs and metal protection systems.

Going from the sublime to the canine, the well-known affection of dogs for lamp-posts causes city lighting engineers some harassing problems owing to the corrosive properties of the liquid with the formula K9P, which eats into the steel of lamp-posts a few inches above the ground. At one time there was talk of investing steel lamp-posts with a small electric charge that would provide a mild shock to persuade dogs to find a more suitable tree. However, the British sense of fair play triumphed and dogs have not been penalized in this way.

To understand corrosion, first consider the chemical action whereby the surface of a metal is attacked by gases in the atmosphere, for example oxygen gradually converting iron into iron oxide. If iron is stored for long periods in contact with dry air it does not rust as much as would be expected from the usual behaviour of the metal. If the iron is kept moist, rusting proceeds rapidly, which might lead one to expect that water was the agent causing the corrosion. But in air-free distilled water rusting does not take place as fast as when air is dissolved in the water. Even the combined presence of air and moisture does not fully account for the formation of rust, for pure iron is attacked less than commercial grades of iron or steel. These observations led to the realization that corrosion is stimulated when a metal is in contact with liquids such as water containing gaseous or solid substances.

Findings such as these confirm that there is a relationship between corrosion and electrolysis. It has been proved that corrosion is accompanied by small localized electric currents. The presence of particles of an impurity, or contact with some other metal, allows a difference of voltage to be set up, causing a minute electric current to flow. The moisture that is present, containing air or some dissolved substance, conducts the electricity and corrosive attack is begun.

Metals and alloys can be listed in what is known as a galvanic series, which helps to predict the likelihood of corrosion of pairs of dissimilar metals in corrosive media. Magnesium and zinc are near the top of the list, showing that they are likely to be preferentially ('sacrificially') corroded when in contact with other metals or alloys. Titanium, nickel–chromium alloys and platinum are near the bottom of the list, showing that they will be protected by the more active ones. North Sea platforms are massively protected by zinc sacrificial anodes welded to the submerged parts of the structure.

When two metals are in contact in conditions in which electrolytic action may take place, corrosion of one metal will occur. This effect, known as galvanic corrosion, was recorded over two hundred years ago. The Royal Navy frigate *Alarm* had her hull covered with copper to prevent damage by barnacles. After only two years' service it was found that the copper sheathing was detached from the hull in many places because the iron nails that had fastened the copper to the timber had become severely corroded. Some nails, which were much less corroded, had been insulated from the copper by brown paper which was trapped under the nails' heads. The copper had been delivered to the dockyard wrapped in this paper, which had not been removed completely when the sheets were nailed to the hull. It was, rightly, concluded that iron should not be allowed to be in direct contact with copper in sea water.

This type of galvanic corrosion has occurred many times because designers and engineers did not realize the havoc it would cause. As recently as 1982, during the Falklands War, two Sea Harriers suffered nose wheel collapse because of galvanic corrosion between the magnesium alloy wheel hubs and their stainless steel bearings.

Sometimes the corrosion behaviour inside a crack or recess in a metal article provides an example of what is known as 'differential aeration'. Moisture which has penetrated to the bottom of the crack becomes devoid of dissolved atmospheric oxygen, while at the mouth of the crack the water contains plenty of oxygen. This provides an electrical potential difference: effectively, an anode at one end and a cathode at the other. Under such conditions the lower end devoid of oxygen becomes the anode, and corrosion takes place.

METHODS OF COMBATING CORROSION

The search for alloys with superior resistance to corrosion is a never-ending attempt to satisfy an insatiable demand.

If the initial coating of corrosion product is adherent and impervious, it can prevent further attack from taking place. Aluminium provides an example of this, for it forms a thin but strong film of aluminium oxide on its surface, which restricts corrosion. Stainless steels containing chromium and sometimes nickel behave even better, their corrosion resistance being provided by the formation of a strong protective oxide film. On the other hand, ordinary iron or steel is liable to progressive corrosion because the layer of rust on the surface is porous and tends to flake off, so further corrosion penetrates inwards, though at a reduced rate.

Protective coatings of other metals can be applied by electroplating, galvanizing, spraying or cladding one metal with another, as when sheets of a strong aluminium alloy are clad with pure aluminium (described on page 160). Most iron and steel is protected by paint, bitumen or proprietary preparations. The Eiffel Tower needs to be repainted every seven years. During the past generation the amount of maintenance on bridges has been much reduced, thanks to a better understanding of the mechanism of corrosion, carefully organized codes of practice for treating steel structures, and the development of improved methods for protecting steel including sprayed coatings of zinc or aluminium. Nevertheless, it is difficult to protect the underneath surfaces of river bridges; the moisture that is present on the steel for most of the year induces the breakdown of coatings and makes it hard to obtain the almost clinical conditions required for optimum performance of the protective treatment.

The use of synthetic resins in paints has increased enormously, thus giving a long life to the paint film; finishes consisting of plastic materials and rubberized paints have also been introduced. Each type of protective coating has its own particular aesthetic or economic merit. One of the most recent developments has been the use of inhibitors, which are chemicals put into cooling and central heating systems and in anti-freeze solutions in car engines to prevent corrosion. Chemicals such as phosphates, silicates, chromates and benzoates are widely used.

The spectacular Thames Barrier, illustrated in *Plate 27*, provides examples of complex corrosion protection problems. The housings for the machinery that rotates the gates are timber structures clad with stainless steel sheeting. The ribs of each housing are fabricated from a hard wood in the form of laminated sheets bonded together. The height of the five larger housings is about 19 metres; standing inside them, one has the impression of being beneath the vaulted ceiling of a cathedral. When the barrier was being designed and corrosion protection was being considered, somebody remarked, 'If it can corrode it will; if it won't, it's gold and somebody's going to pinch it.' Eventually it was decided that, though stainless steel was not quite so effective as gold, it would be a good substitute.

Each wooden structure supports a shell roof fabricated from stainless steel containing about 17 per cent chromium, 12 per cent nickel and 3 per cent molybdenum. The River Thames has been cleaned up, so that fish can once more live in it, but this means that there is a greater concentration of oxygen in the water, and thus an increased tendency to corrosion. The stainless steel shells are only decorative covers, so the minor corrosion problems which will be encountered should not affect the barrier engines, but eventually a 'facelift' will be required. The other steels used in the construction of the gate and gate arms are given a pre-treatment by projecting fine grit at high velocity at the surface, after which they are covered with a blend of epoxy resin and coal-tar pitch.

CATHODIC PROTECTION

It is sometimes possible to reduce corrosion by a method called cathodic protection, illustrated by the galvanizing of steel. The zinc coating slowly corrodes in preference to the steel, whose corrosion is thus delayed; this effect is also described as sacrificial protection. Several methods of applying cathodic protection exist, but they all work only in an electrolyte which would otherwise cause corrosion, for example sea water, polluted waters and moist soils. They cannot be used as an antidote to atmospheric corrosion.

Sacrificial blocks of zinc or aluminium, usually alloyed to ensure maximum efficiency, can be attached to the structure to be protected, either through a cast-in steel insert or an insulated cable. Ships, offshore oil exploration and production platforms, jetties and submarine pipelines are protected in this manner; so too are many steel pipelines buried on land, especially those carrying gas, oil or water under high pressure. Magnesium anodes are often used for protection in such conditions.

In some cases it is possible to eliminate the need for a sacrificial anode and replace it with a permanent inert anode of graphite, lead, iron–silicon alloy or titanium coated with platinum. In such a system the electrochemical effect which prevents corrosion is derived from a low voltage source of direct current, transformed and rectified from the mains or even from a solar cell system. These may be used where a source of power is available. In either method it is common to have a coating as a first line of defence against corrosion, with cathodic protection ensuring that corrosion does not occur where the coating has become porous or damaged.

STRESS-CORROSION

Failures due to stress–corrosion cracking or corrosion fatigue take place at stresses well below the normal tensile limit of the metal involved. As the names imply, in the first case fracture results from the combined influence of corrosion and tensile stress, while in the second case rapidly alternating stresses under corrosive conditions, as for example in an aero engine, lead to failure.

For a given metal system there is usually a specific corrosive agent or series of agents capable of initiating cracking. Thus stress corrosion in brass takes place in the presence of ammonia, while for stainless steel it is the presence of chlorides that is the hazard. For mild steel, contact with strong caustic soda can result in 'caustic cracking' which sometimes affects the rivets of high pressure steam boilers. The caustic agent comes from alkaline waters, which can seep into and concentrate in the crevice under the rivet head. There seems little doubt that the initiation of cracking results from a process of corrosion operating on a microscopic scale.

Stress–corrosion failures fall into two categories: those taking place in the presence of stresses introduced into the metal during a deformation process and remaining 'locked up', and those stresses imposed by external forces applied during the assembly of equipment or during its operation.

Stress–corrosion from locked-up stresses caused accidents to British infantry and cavalry in India during the late nineteenth century. In the monsoon season, ammunition was stored in stables until the dry weather returned, but by then many brass cartridges were found to have suffered what became known as 'season cracking', which led to bad accidents when cartridges 'blew back' when fired. The cracking seemed to have been caused by some form of corrosion, though the actual attack on the metal was only slight. Other examples of season cracking were noticed, and it was remarked that they occurred only when a metal had been cold worked in its fabrication and later subjected to corrosive conditions. It was not until 1921 that the cause of season cracking was discovered. The cartridges had been subjected to stresses in their manufacture (see page 171). Corrosive conditions were provided by the combination of humidity in the atmosphere plus horse urine in the stables. When the mystery was solved, it became possible to remove the pent-up stresses in the brass by annealing at a temperature of 200 to 300°C. By this treatment the stress is relieved, so that the brass will not tend to season crack, and the annealing temperature is not sufficiently high to reduce the mechanical strength of the metal.

CONDENSERITIS

Our ships, great and small, have been at sea more continually than was ever done or dreamed of in any previous war since the introduction of steam. Their steaming capacity and the trustworthiness of their machinery are marvellous to me, because the last time I was here one always expected a regular stream of lame ducks from the fleets to the dockyards, with what is called 'condenseritis' . . . but now they seem to steam on almost for ever.

These words were spoken by Winston Churchill when he presented the Navy Estimates to the House of Commons on 27 February 1940, but probably few appreciated that his remarks paid tribute to years of painstaking research work by a handful of British firms. In all forms of steamship, a condenser is vital to secure maximum efficiency; it consists of nests of tubes about 4 to 7 m long and 18 mm in diameter, with walls about 1.5 mm thick, all arranged in a gigantic honeycomb. Cold sea water is circulated through the condenser tubes while, outside the tubes, exhaust steam from the engines condenses to pure water and is returned to the boiler for generating more steam. If the tubes corrode, the sea water leaks into the boiler water, causing rapid corrosion of the whole power unit.

During the First World War the average life of condenser tubes, which were then usually made of 70/30 brass, was only a few months, and in the early years of the Second World War many naval vessels were immobilized because of failures in the condenser tubes. In 1940 the developments mentioned by Winston Churchill were taking place. At first a copper alloy with 5 per cent nickel and 1 per cent iron gave a much improved condenser life, and it had the advantage that existing copper-working techniques could be used. During the following years the flow velocity of the cooling water was nearly doubled, and the 5 per cent alloy could not resist the eroding effects of the water's turbulence. A copper alloy with 10 per cent nickel was introduced for surface vessels and one with 30 per cent nickel for submarines.

The number of steamers declined in favour of oil-fired ships, also requiring condensers that would withstand severe corrosion. Then desalination plants and the equipment of the offshore oil industry demanded solutions to further problems. Initially oil platforms in the North Sea had galvanized steel for piping systems conveying sea water, but rapid corrosion led to expensive replacement and repairs. Then a copper alloy with 10 per cent nickel was introduced and gave a greatly improved life. It cost about a third more than galvanized steel, but that was insignificant compared with the savings in downtime and repairs.

During the 1970s, the British electricity generating boards stipulated that all new coastal power stations, and especially nuclear stations, should have titanium condenser tubes. A typical installation has about 20 000 tubes, each more than 20 m long.

EROSION AND CAVITATION

Some parts of marine propellers, turbines, hydraulic and hydroelectric pipelines suffer from cavitation attack, caused by the breakdown of streamline flow in the liquid passing over the surface of the metal. Sudden changes in flow produce 'vacuum bubbles'; where these bubbles collapse, the metal surface is damaged. Metal tubes carrying water may suffer high water velocities from eddy currents, which will erode the protective oxide film on the metal. On return to quiescent conditions the eroded part re-oxidizes, and this process may be repeated several times until the tube wall is penetrated. Characteristic 'horseshoe' shapes are developed in these corrosion–erosion areas. The solution of the problem is normally to replace the joint or obstruction which has been causing the eddy currents. Resistance to this attack can also be improved by adding between 1.5 and 3.0 per cent iron to the copper–nickel alloy.

MICROBIOLOGICAL CORROSION

A micro-organism can be viewed as a catalyst which can produce high concentrations of hydrogen sulphide or organic acids from materials containing sulphate ions or petroleum hydrocarbons. The most important group of bacteria associated with the corrosion of metals is the sulphate-reducing bacteria. They obtain their carbon and energy from organic nutrients, and their respiration makes use of sulphate ions, which results in the formation of corrosive sulphides.

MARINE FOULING

Marine organisms can attach themselves to condenser tubes, blocking generator intake ducts and filter screens. One solution is to back-flush the system with chlorine, but a more environment-friendly way is to use copper alloys, which are toxic to marine organisms. Hence copper alloyed with 10 per cent nickel is used to line seawater ducts. This anti-fouling property is used to good effect in fish farms, where the cages containing the fish are of copper–nickel alloy mesh.

Marine fouling of offshore structures by weed and shellfish causes increased stresses because of their considerable weight, and by the increased drag they cause in tidal currents. Cladding the legs of the platforms with copper containing 10 per cent nickel prevents the build-up of fouling. A fleet of shrimp boats known as the Copper Mariners are protected with the alloy, and have never needed to be dry-docked in twenty years of service in the Caribbean.

CORROSION TESTS

There are an enormous number of corrosion conditions, each one presenting individual problems. It has been reported that over a thousand different fluids are transported by tankers, and for each of them it is necessary to test the material that is to contain the liquid. The behaviour of metals and alloys in service can be forecast by subjecting them to accelerated corrosion tests. As a result of years of experience, these tests give fairly accurate results, provided control factors are taken into account. Such corrosion tests are usually conducted in closed chambers with water sprays containing sodium chloride, hydrogen peroxide or acids, whichever causes the appropriate type of corrosion. In another type of test, the specimen is alternately sprayed to wet it, and dried by a stream of warm air.

In some corrosion tests, liquids are directed at high speed onto portions of tube or flat specimens which are strung together and electrically insulated from each other. The specimens are weighed before and after testing; they are examined to determine the type of attack, whether general thinning or localized. In other tests, jets of liquid are fired at samples of various metals and alloys. As a result of many such tests, it has been found that the state of flow of a liquid, whether it is streamlined or turbulent, can have a critical effect on corrosion.

Corrosion tests therefore help to predict when corrosion is likely to occur, but they should be regarded as only one of the weapons needed in the everyday fight against corrosion. Nowadays the chief concern of corrosion scientists is to 'get the message across' to the architects and designers on the job, so that they will take advantage of all the knowledge that has been obtained on the prevention of corrosion. It is rather striking that the growth of knowledge about this complex subject has caused an important change of attitude. At first, corrosion was regarded as an unavoidable evil, and the best that could be done was to build structures with more than ample weight of metal and apply paint liberally and regularly. Now metallurgists, corrosion specialists, designers and engineers have gone on the attack, with the object of preventing corrosion from taking place. The fight against corrosion starts on the drawing-board.

24

JOINING METALS

Some components are designed in one piece and manufactured by processes such as diecasting or investment casting. Other items, perhaps more complex in shape, can be made effectively by assembling several parts of simpler design. The attention which has been paid in recent years to processes such as soldering, brazing, welding, and joining by adhesives has made 'built up' construction rapid and efficient and often cheaper than 'one piece' manufacture. There has been spectacular progress in the joining and assembling of metal articles, from the manufacture of bicycles to the building of ships.

MECHANICAL METHODS

The Bayeux Tapestry shows a fine display of armour, and illustrates some ancient methods of joining metals. The sheet iron helmets were joined by riveting to the 'nasals' which protected the warriors' noses. The armour worn at the Battle of Hastings was fashioned by solid welded rings alternating with riveted rings.

During and after the Industrial Revolution, the process of riveting was widely and noisily used, from Brunel's *Great Britain* to the *Queen Mary* nearly a century later. A number of well-known structures provide other examples of the use of riveting, from the Statue of Liberty with a mere 300 000, to the Eiffel Tower with $2\frac{1}{2}$ million, and the Forth rail bridge with 8 million rivets.

SOLDERING AND BRAZING

The operation of soldering consists of uniting metal components by an alloy which melts at a lower temperature than either piece of metal to be joined. When liquefied, the solder covers the heated surfaces and forms a thin alloy layer, so that when the parts have cooled the two pieces of metal remain firmly united. Metals which are to be soldered must be cleaned before commencing work, and in order to maintain the clean surface and remove oxide films, a flux such as zinc chloride is usually applied. The most familiar solders are those containing lead and tin. The solder is melted with a blow-lamp or heated soldering iron (the 'iron' actually being of copper).

The ease with which a solder will wet and flow over the surface of another metal depends on the characteristics of that metal. For example, molten lead will not wet a copper surface, but if a little tin is added to the lead the resultant alloy will readily flow over copper. Aluminium used to be regarded as a difficult metal to solder because the tenacious film of aluminium oxide which forms on the surface interferes with wetting. Similar problems arise with alloys containing aluminium; as little as 0.5 per cent aluminium makes soldering somewhat difficult. Special fluxes have been developed for soldering aluminium and, although some care is necessary, satisfactory joints can be made. Considerable use has been made of this development in the electrical industry, for making joints in aluminium wire cables.

Silver solders are alloys of copper, zinc and silver, sometimes with small additions of cadmium. One solder contains 60 per cent silver with 30 per cent copper and 10 per cent zinc, and melts at 735°C. At the beginning of this century silver solders were used mainly for jewellery, but later they were introduced for engineering use on account of their strength, fairly high melting points, and the accuracy and neatness of the joint. The frames of some racing bicycles are joined with silver solder.

For many years, bicycle frames have been made by joining steel tubes to brackets by brazing. The brazing solder is a form of brass (hence the word 'brazing') which consists usually of 60 per cent zinc with 40 per cent copper. The alloy can be melted by a gas torch. A brazed joint is considerably stronger than a soft-soldered one, and partly for this reason there has been a tendency towards brazing instead of soldering, a trend that received further impetus from the rise in the cost of tin. The frame of the modern ordinary domestic bike is made from fairly substantial steel tube, so nowadays the frames are welded by using the MIG process, which will be described on page 240. This lends itself well to automation. The frames of racing and mountain bikes are joined by brazing silver soldering or adhesives, which add less weight to the structure than weld metal.

Many aluminium alloys can be brazed, and the development of aluminium radiators for automobiles depended on the intense development work that made high-speed brazing of aluminium possible. The filler metals are aluminium–silicon alloys with from 7.5 to 12 per cent silicon. As with soldering, chemical fluxes are required, except where the brazing is carried out in vacuum or inert gas.

WELDING

Some metals will weld even without heat being applied; for instance, two clean flat pieces of lead, silver, gold or platinum can be laid face to face and hammered cold until they become welded. Most other metals and alloys have a temperature above which hammer welding becomes possible, though oxidation of the surface of the metal may interfere with the success of welding at this minimum temperature. Aluminium is difficult to hammer weld because the thin layer surface of oxide prevents metal-to-metal contact. Cast iron is practically impossible to hammer weld.

Chains used to be made by strenuous manual effort: each piece of hot iron bar was threaded through the preceding link, turned over at the ends and made into a link by hammer welding.* Nowadays, chains with links up to 40 mm in diameter are made automatically, the round bars being formed and bent to shape and then welded in a second automatic operation. Chain links of over that size are shaped on hydraulic bending machines and then welded.

The medieval sword-maker had to be an expert in hammer welding. He was faced with the problem of combining the sharp cutting edge obtained in a high carbon steel with the need to prevent the sword from shattering if it was too brittle; but if the sword was too soft it would bend and be equally useless. The solution was to hammer layers of metal together, a soft iron or low carbon steel in the centre and a much harder high carbon steel to provide the cutting edge. All the pieces would be hammer welded together to make a sword blade with the required properties, then the weapon would be fashioned and decorated. In the third century a process now called pattern welding was used in which iron rods and wires were twisted together, heated and forged so as to join them by welding, and some very decorative swords were made in this way.

---

* The Black Country Museum at Dudley in the UK has numerous fascinating exhibits of 'old-time' industries, including a demonstration of chain-making.

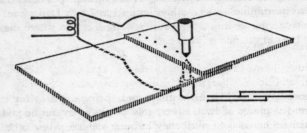

Fig. 43. Spot welding

ELECTRIC RESISTANCE WELDING

When an electric current is passed through two pieces of metal in close contact they heat up, depending on their electrical resistance; heat is also generated at the points of contact. This is applied in a process known as resistance butt welding. The two metal parts are placed in separate electrodes and butted together while a high amperage electric current at a low voltage is passed across the joint. Heat is developed at the point of contact and when the welding temperature is reached, pressure is increased and a joint is formed between the two components.

Another method, known as spot welding, illustrated in *Fig. 43*, is particularly suitable for joining sheets of metal which are held together with two electrodes lightly pressing on either side. An electric current is passed through the electrodes, thus heating the metal between them. Immediately afterwards the two electrodes are pressed tightly together, and the metal sheets are joined effectively. Spot welding is used for joining motor vehicle bodies, aircraft, and other sheet steel fabrications such as washing machines and refrigerator cabinets.

A further development of spot welding is seam welding, in which a pair of metal sheets is passed between electrodes in the form of rollers, so that a continuous line of weld is obtained.

Welding electrodes must have good electrical conductivity and are usually of a copper alloy such as copper–chromium, copper–chromium–zirconium, or occasionally copper–cobalt–beryllium or copper–nickel–phosphorus. The electrodes need maintenance after about ten thousand welds, and then they may be refaced.

In ultrasonic welding two metals, not necessarily of similar composition, are placed in close contact and subjected to high frequency vibrations. Special equipment is required to convert electrical energy into rapid mechanical vibrations, and the process is used for bonding lead-out wires for semiconductor devices and for encapsulating explosives.

OXY-FUEL WELDING

For welding joints that are larger than those that can be tackled by spot welding, additional filler metal must be introduced between the two pieces to be joined. The filler metal and adjacent parts of the parent metal are melted and allowed to fuse together, thus uniting the two metal components. Oxy-fuel gas welding can be done with propane or butane, but acetylene has advantages over those gases because it provides a higher flame temperature, about 3000°C, and uses less oxygen (only one-third of that required for propane).

By adjusting the supply of oxygen or acetylene from the gas cylinders, the type of flame can be regulated according to the requirements of the metal to be welded. Some metals, for example brass, weld better if the amount of oxygen is in excess of that required to burn acetylene, while cast iron welds better in a reducing flame in which surplus acetylene is present.

The flame is played over the parts of the metal to be joined, and also a metal filler rod which melts and runs into the gap between the two metal parts and joins them. A flux has to be used in welding aluminium, though it is not necessary for steel.

ARC WELDING

The metallic arc process is used to weld ships' structures, large machines, structural steelwork and pipelines. The heat is developed by striking an electric arc between the work and a metallic filler rod. This is usually of a low carbon steel, but additions of metallic powders and deoxidizers to the flux coating modify the composition of the weld metal. The formation of the arc develops so much heat that the end of the rod melts and forms a weld. The rod is coated with suitable flux which melts when the arc is struck, so that the weld solidifies and cools under the protective coating of molten flux.

There are several variations of arc welding. In the tungsten inert gas (TIG) process, formerly called argon arc welding, an arc is struck between the component to be welded and a tungsten electrode held in a suitable container around which flows a stream of inert gas, usually argon. This protects the heated tungsten and the welded metal from oxidation, thus dispensing with the use of fluxes. TIG welding has revolutionized the joining of non-ferrous metals and has made it easy to join aluminium, which used to be difficult to weld because of the oxide skin which so readily forms on the metal. It has also become possible to weld titanium, zirconium and tantalum, thus helping to widen the use of those metals in aircraft, chemical engineering and other industries. The TIG process is used for welding stainless steel in nuclear reactors and chemical plant.

When welding any aluminium alloy, an alternating current arc is necessary. The arc exerts a cleaning action on the surface of the aluminium which removes the thin tenacious film of aluminium oxide and enables the edges of the materials to melt and flow together. With TIG welding there is virtually no limit to the thickness of material that can be joined, but the process is quite slow and labour intensive. It is popular for the preliminary welding, known as the 'root run', of *in situ* pipes, where the workpiece cannot be rotated. After TIG deposition of the root run, a cheaper process such as metal arc welding is often used to complete the weld. Indeed, some TIG machines can be changed to metal arc at the touch of a switch.

In a variation of TIG, the metal inert gas (MIG) process maintains an arc between the end of a base wire electrode, usually copper-coated mild steel, and the components to be welded. The wire is fed automatically at constant speed, and the weld is shielded by an inert gas, such as argon, or carbon dioxide, or a mixture of the two.

The submerged arc process is used when high productivity is required. The electric arc is maintained between the end of a bare wire electrode and the work. As the electrode is melted, it is fed into the arc by a set of rolls driven by a motor. The arc operates under a layer of flux (hence 'submerged' arc).

The various techniques of arc welding offer rapid and effective methods of construction in shipbuilding. By the use of welding instead of riveting, weight is saved, since riveted joints require the ship's plates to be overlapped, and each rivet adds its little load to the dead weight of the vessel. Large and small naval vessels have been made with extensive use of arc welding. Ten million rivets were required in the *Queen Mary*, built in the 1930s, but thirty years later the only rivets in the *Queen Elizabeth 2* were those used to connect the aluminium superstructure to the hull.

NEW DEVELOPMENTS

In electron beam welding, the work-piece is heated by the bombardment of a beam of electrons. Filler metal may be added if necessary. The work is usually done in a vacuum chamber, but electron guns are being developed which allow the electrons to travel a short distance through air. This process is used for dealing with difficult-to-weld materials, and for such specialized products as turbine-blade root joints. Laser welding will be discussed on page 272.

Diffusion bonding is used to weld heavy section constructions, hollow assemblies and other inaccessible joints. The parts to be joined are machined, cleaned and clamped together, then placed in a vacuum chamber and heated. Bonding normally occurs at about 70 per cent of the melting temperature of the material (in degrees Celsius); a diffusion bonded joint is produced, with good mechanical properties.

Magnetically impelled arc butt welding, known as MIAB, is used in the mass production of tubular assemblies such as automotive back axles. Two steel tubes are clamped so that there is a gap of about 2 mm between them. One is advanced and retracted to form an arc, which is rotated round the joint interfaces by the interaction of the magnetic fields set up by the coils and the arc current. By the conclusion of the timed cycle, about 6 seconds, the ends have become joined together.

A development which has grown from an exotic laboratory technique in the early 1970s to a present-day practical manufacturing tool is known as plasma arc or plasma jet welding. In this type of welding high energy inputs are possible, up to 50 kW, and alloys can be fed in as powder. Materials such as borides, carbides and nitrides, which have very high melting points, can be melted and sprayed onto a backing material such as steel. An example of the capability of this process is the 30 km of plasma welding that was carried out to line concrete tanks with stainless steel sheet for a copper extraction plant in Zambia. More recently, the plasma arc process has been extended by what is known as plasma-MIG welding, a combination of plasma arc and argon arc welding.

The friction welding process that was developed about thirty years ago is now being used by all industrialized countries. One part is rotated at high speed and then pushed hard against another, causing the two to be welded. Such friction welds can be made between dissimilar metals: a typical example is provided by the aluminium conductors used in aluminium refineries. The conductors cannot withstand the very high temperatures close to the furnace, so at that point steel is used, friction welded to the aluminium.

Fully automated welding by robots is already with us. Fiat in Italy were pioneers, and a production line in Turin features the assembly of car bodies by means of automatic spot welding. Another impressive achievement is in the shipyards of Japan at the Ariake yard of Hitachi, where more than three-quarters of the fabrication of supertankers is by means of automatic welding. *Plate 28b* shows automatic MIG welding by a robot.

The latest developments within the welding industry have aimed at increasing productivity and quality by improving the power source response, accuracy and control. Such improvements have led to the increasing adoption of mechanized, automated and robotic welding in many industries. A completed weld, whether deposited by human or by machine, is dependent on the skill of the welder, or the programming of the machine and expertise of the welding engineer. A weld has been described as the whole of metallurgy in miniature, for it calls for a knowledge of the melting, casting and forging of metals. A weld carried out by an inexperienced operator, or by an incorrectly programmed robot, may be weak, brittle, coarse grained and unsound. On the other hand, skilled operating and planning can produce welds that are at least as strong as the parent metal. The weld can be proved to be satisfactory and reliable by inspection techniques such as radiography and ultrasonic testing.

UNDERSEA WELDING

The task of undersea welding of oil platforms, bracings and pipelines offers serious challenges to the ingenuity of welders and welding engineers. Three basic techniques have been adopted. The simplest and cheapest is 'wet' welding, which requires no specialized equipment and generally involves metal arc, metal inert gas or flux cored arc welding. However, high quench rates due to the surrounding water coupled with the pick-up of hydrogen into the weld may lead to poor quality welds of inadequate strength and toughness.

In hyperbaric welding, illustrated in *Fig. 44*, a chamber is built round the joint to be welded, and water is displaced from the chamber by gas pressure. The pressure inside the chamber is therefore equal to that prevailing outside at the depth of the chamber. Hyperbaric welding has become an accepted method of underwater repair and maintenance welding, down to depths of about 300 metres. At depths greater than this, problems are experienced because of the high pressure inside the chamber; replacement of human welders with automated or robotic equipment is one possible solution.

Atmospheric welding can be performed at depths below which saturation diving is not possible. In this system a working pressure equivalent to the atmosphere is maintained. The system is based on the use of a

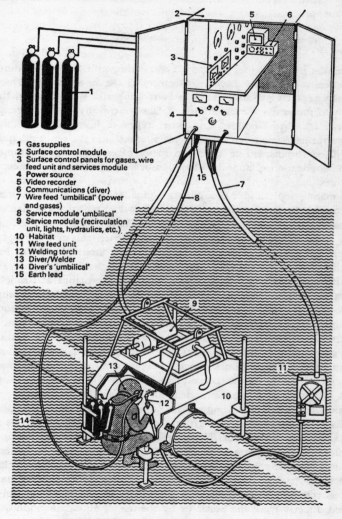

1 Gas supplies
2 Surface control module
3 Surface control panels for gases, wire
   feed unit and services module
4 Power source
5 Video recorder
6 Communications (diver)
7 Wire feed 'umbilical' (power
   and gases)
8 Service module 'umbilical'
9 Service module (recirculation
   unit, lights, hydraulics, etc.)
10 Habitat
11 Wire feed unit
12 Welding torch
13 Diver/Welder
14 Diver's 'umbilical'
15 Earth lead

Fig. 44. Undersea welding (*Courtesy BOC*)

disposable pressure vessel or work chamber, linked to a second, reusable diving sphere that contains all the equipment necessary for welding and its related processes. This is the most costly of the three methods.

ENGINEERING ADHESIVES

During the last fifty years the use of engineering adhesives has increased dramatically. Not only have they improved the appearance of metal items by replacing techniques such as riveting and spot welding, they have also increased the reliability of components by adding strength to threaded, shrink-bonded and other types of joint. The adhesives themselves have improved over the years, and so has the range of equipment used for dispensing them. In many cases automatic equipment for the application of adhesives has significantly speeded up production processes.

To many people a glue is a glue is a glue. However, there are many different kinds of adhesive available, each with its own particular use. For example, there are industrial epoxies. These differ from the traditional DIY products in that they are toughened through the use of rubber, and have just a single component (the household versions have two that need to be mixed). The benefit of toughened epoxies is that they have greater peel strength and impact performance at the joint. Next come the toughened acrylics. These are two-part adhesives which develop their strength at room temperature within 15 to 30 minutes, although some can be subsequently heat cured.

Phenolic adhesives are solvent based and are used for applications such as bonding brake shoe pads and clutch discs. Cyanoacrylates – instant adhesives (superglues to many people) – cure in seconds at room temperature. This curing takes place through the reaction of slight traces of moisture on the component surface with the adhesive. Instant adhesives are ideal for bonding rubber, some fabrics and most plastics.

One of the most important groups of adhesives, as far as metals are concerned, is the anaerobics. These are single-component materials which remain liquid in the presence of air and only cure when the joint has been mated. Anaerobic adhesives are used on metallic components for locking and sealing threads, for providing gaskets and for bonding components such as retaining gears, pulleys and bearings on shafts. They can be applied directly from the bottle or by the use of automatic application equipment. The bonded object is strong enough to handle within a few minutes, with a full cure taking between 3 and 12 hours, depending on the thickness of the bond and the temperature of the surroundings.

The ideal time to select the most suitable adhesive is at the design stage, when the materials are being discussed and the shape and size of components are being planned. The assembly process time, the equipment to be purchased and the number of operators necessary are all vital factors in calculating the cost of replacing a previous joining method with adhesives. As with all newly adopted processes, the potential user must obtain advice from the adhesive supplier in order to get the best out of the process.

There are many examples where the use of adhesives has improved the quality of the join while reducing assembly cost. Watch glasses are now sealed into their cases; bicycle frames are adhesive bonded; and a fire engine was reduced in weight by $1\frac{1}{2}$ tonnes by replacing the conventional timber framework and cladding panels with aluminium, fixed together by adhesive rather than nuts and bolts.

25

POWDER METALLURGY

Powder metallurgy is an economical way of shaping components, first by compacting the metal in the form of powder, followed by sintering to increase the density and thus the strength of the component. The process is suitable for metals with very high melting points which would be difficult or impossible to shape by casting or forging. Also, it is a method of shaping which is comparable in cost to other possible processes.

The stringent demands of modern technologies can sometimes be met only by using metals such as tungsten, molybdenum, tantalum, niobium or rhenium, which have very high melting points. They have great strength at high temperatures and great resistance to corrosion, but are extremely difficult to cast. Processing these metals, therefore, is based on powder metallurgy.

Non-metallic materials and intermetallic compounds, which cannot be incorporated in metals by casting, can be embodied in the powder mix. Tungsten carbide blended with cobalt for cutting tools is shaped by powder metallurgy; the production of tungsten carbide cutting tools represents an immense part of the powder metallurgy field. Graphite, an excellent lubricant, can be put into bronze for bearings; carbon is incorporated into copper for current-carrying bushes in electric motors.

Metal powders can be mixed in the proportion needed to form alloys; for example, appropriate amounts of copper and tin powders can be compacted and sintered to form bronze. Lubricating oil and graphite may be added to make an oil-retaining bearing, widely used in the automotive industry, but also featuring in many domestic articles such as washing machines, vacuum cleaners and electric clocks. In another type of blending, friction materials for brake linings and clutch facings are embedded in copper and other materials.

COMPACTING THE POWDER

If necessary, the metal powder is first blended with the material with which it will be associated. Next, the powder, mixed with a small amount of lubricant, is loaded into a die containing the shape of the component, into which it will be compacted at high pressure. High speed steel or tungsten carbide are the usual die materials. The die life depends on the size and complexity of the component, but typical figures are 200 000 'compacts' for a steel die, and about a million for a tungsten carbide die. The dies are mounted on machines which can provide a pressure of 150–700 N/mm^2. The powder is compressed into the die cavity, and the die is opened and the moulding ejected; the compacted part is weak at this stage, and must be handled carefully.

SINTERING

The final process, sintering, is a key part of the operation, as it gives the friable compacted powder its strength. The sintering is generally carried out in electrically heated furnaces at a high temperature, usually at about two-thirds of the melting point of the metal (in degrees Celsius). The furnace has a protective atmosphere of hydrogen or hydrocarbon gases, to which nitrogen is sometimes added. High speed steels and some other alloys are sintered in vacuum.

ISOSTATIC PRESSING

When powders are compacted in dies as described above, there are design limitations. Complex shapes cannot be achieved in the direction of the pressing operation, because the powders are compressed in that direction and there is no lateral flow. Separate punches are therefore necessary to produce features at right angles to the direction of pressing. Isostatic pressing techniques remove the requirement for a rigid die by sealing the powder in a flexible container and applying a uniform hydrostatic pressure over the whole surface of the container through a fluid medium within a cylinder.

In cold isostatic pressing, the powder is encapsulated in a mould of rubber or plastic immersed in water which is then pumped to a high pressure, about 400 N/mm^2. The powder is compacted and the process is usually automated. Hot isostatic pressing is used to compact alloys that are difficult to press or sinter. The container is usually of metal because the process takes place at high temperature. Pressure is applied by argon, and needs to be only one-quarter of that required in the cold process. Hot isostatic pressing is used for the moulding of hard metals such as those used for extrusion dies and for billets of high speed steel, titanium and superalloys.

SINTERED CUTTING TOOLS

For years, metallurgists and engineers sought tools that would be superior to high speed steels for cutting and machining metals. It was found that when tungsten powder was heated with carbon to a temperature of about 1500°C, a compound, tungsten carbide, was formed; this is exceedingly hard and very suitable for making cutting tool tips. Tungsten carbide powder is blended with 6 to 10 per cent cobalt. This mixture is pressed into blocks and heated in hydrogen to a temperature of about 1000°C; this pre-sintering makes a product which, though hard, can still be machined and ground to the required shape and size. The tool tips are then given a final sintering at about 1400°C, which makes them much harder than the hardest steel. The tips are brazed or mechanically locked on the end of steel shanks. The great hardness of tungsten carbide has produced drills that are able, for example, to drill holes in glass. Tungsten carbide is used for making dies or die inserts for such processes as extrusion and wire drawing, and for the compacting tools that produce other powder metal components. A variety of intermetallic compounds can be added to give even greater hardness. Cobalt has already been mentioned; titanium carbide or tantalum carbide are also added.

So-called 'oil-less' or 'self-lubricating' bearings are made by sintering a mixture of copper and tin powders with graphite. They are then soaked in oil, and no oil holes or grooves are required since oil will soak through such a bearing as though it were a sponge. In some cases, the amount of oil held in the bearing is enough to last the lifetime of the machine, and some are so porous that they could be used as wicks in oil lamps. Self-lubricating bearings are used in the automobile industry, and in many domestic articles such as washing machines, vacuum cleaners and electric clocks.

Many refractory metals, including tantalum, molybdenum and niobium, as well as some other metals such as nickel, copper, cobalt and iron, can be extracted from their ores to yield the metal in the form of powder. Hence there is an incentive to fabricate by powder metallurgy methods; 'ingots' of molybdenum and of nickel weighing a tonne or more have been made by pressing and sintering.

Powder metallurgy is a comparatively young industry. After a period in which it found its feet, it moved ahead with considerable verve. In addition to the obvious advantages of producing a finished component that does not require further machining, it offers some advantages relevant to today's problems. Process scrap is very small, amounting to only about 1 per cent. This can be compared with the foundry industry where the surplus metal – the runners and risers – may amount to a third or more of the weight of the finished component.

26

METALS AND NUCLEAR ENERGY

Atoms are very small and very numerous; in a piece of metal the size of a pinhead there are many more atoms than the total of all the letters in all the books that Penguin Books has ever published. Each atom contains a central nucleus which is positively charged; negatively charged electrons orbit the nucleus but are prevented from uniting with it by their speed in orbit, just as the Earth is prevented from falling into the Sun. The atom of hydrogen has one electron circling a nucleus of one positive charge, the proton. Helium has two electrons round a nucleus containing two protons; lithium three, iron twenty-six and uranium ninety-two. The number of protons in the atom of an element is known as its atomic number.

Electrons have only about one eighteen-hundredth the mass of protons; they behave as though they move around their nucleus in one or more rings. Hydrogen and helium have only one ring; lithium's three electrons are arranged in two rings; aluminium has three rings for its thirteen electrons; iron, nickel, copper and zinc have their electrons in four rings; some of the heavy metals including silver and molybdenum have five rings; gold and the rare earth metals have six rings. Uranium and radium have seven rings. The outer-ring electrons are principally responsible for the chemical properties of each element.

ISOTOPES

When relative atomic weights* were first calculated from chemical experiments, surprisingly they were not whole integers; for example, chlorine was found to have an atomic weight of $35\frac{1}{2}$. The explanation came in 1913, from a British scientist, Frederick Soddy, who deduced the existence of isotopes – atoms of the same element, differing in atomic weight

* Relative to hydrogen, which was counted as having an atomic weight of unity.

but with identical chemical properties and identical physical properties, except those determined by the mass and behaviour of the isotopes. In 1919 Frederick Aston was able to prove that about three-quarters of chlorine has an atomic weight of 35, while the other quarter has an atomic weight of 37; this mixture gives the actual atomic weight of 35½ mentioned above. Since the chemical properties of the two isotopes of chlorine are identical, they cannot be separated by any chemical process, but Aston isolated them in minute amounts by using an instrument called the mass spectrograph. Under an appropriate electromagnetic field, the two isotopes can be made to follow slightly different paths, because of their differing masses; they can then be separated in the same way as two rockets of differing weights fired off with a similar thrust will follow differing trajectories.

Although most elements are a mixture of two or more isotopes and sometimes (as with tin, for example) as many as ten, one isotope is generally present in a preponderant amount. Thus oxygen contains 99.80 per cent of oxygen-16, 0.03 per cent of oxygen-17 and 0.17 per cent of oxygen-18. Even hydrogen was found to contain 0.02 per cent of an isotope with atomic weight 2, which became known as 'heavy hydrogen' or 'deuterium' (symbol D). 'Heavy water', formed by uniting heavy hydrogen with oxygen, has important uses in nuclear energy research, and in some nuclear power stations. It is usually given the symbol D_2O, to distinguish it from H_2O.

Each element has a characteristic number of protons, the positively charged particles; the nucleus also contains neutrons, which are uncharged particles having very nearly the same weight as a proton. Electrons, equal in number to the protons and each with a negative electric charge counterbalancing the positive charge of a proton, circle round the nucleus in one or more rings as described above. The presence of the neutrons does not affect the chemical properties of the element, but only changes the mass of the nucleus. In a given element all the isotopes have the same number of electrons, which provide its characteristic chemical properties. They have the same number of protons in the nucleus, but each isotope has an individual number of neutrons. It was discovered later that different isotopes of some metals, including uranium, possessed very individual properties in their radioactive behaviour and in their response to 'fission'.

SPLITTING THE ATOM

Early this century, Ernest Rutherford and his co-workers realized that it should be possible to break away some protons from the nucleus of an atom. The most suitable method then available was to bombard the nucleus with a stream of the 'alpha particles' which are spontaneously thrown off from radioactive elements such as radium. They are the nuclei of helium atoms which have lost their orbiting electrons and so have a net positive charge. By directing a beam of these alpha particles onto thin layers of various elements, Rutherford was able to witness scintillations on a fluorescent screen, indicating that a nucleus had been hit, but these collisions were only achieved atom by atom, and thus the mass affected was infinitesimal. In the course of these experiments it was suggested that if the atom could be 'split', a valuable source of energy might be obtained; indeed, in the 1930s many ordinary people were confident that the splitting of the atom would solve all our energy problems.

The problem was twofold: the targets were very small (if a nucleus can be said to have a 'diameter', it is only one ten-thousandth the 'diameter' of an atom); and a positively charged particle used as a projectile would have difficulty in hitting its minute target, because positive charges repel each other. When Rutherford was pressed to give a prophecy about the atom as a source of energy, he remarked, 'Anyone who looks for a source of power in the transformation of the atom is talking moonshine.'

The discovery of the uncharged neutrons in the nucleus provided a type of particle which would not be repelled by the protons in the nucleus, but the problem remained depressingly elusive because of the atom-by-atom approach that was necessary. What was needed, but what did not then exist, was a method by which, when one atom was split, the disintegration separated further neutrons, thus causing what was later to become known as a 'chain reaction'. If, for example, two neutrons could be given off, they might cause two other atoms to be split, which would yield four neutrons, then leading to 8, 16, 32, 64, 128, and so on in rapidly increasing amounts.

During pioneering experiments in neutron bombardment, many materials were changed, in minute quantity, to new isotopes or new elements. The mechanism of this process may be understood by considering an atom of carbon, which consists of six protons and six neutrons in the nucleus and six orbiting electrons. If the carbon atom is bombarded by neutrons, it is converted into an isotope with six protons and *seven* neutrons, still with six electrons. This is a stable isotope, 'carbon-13', which is contained in natural carbon to the extent of about 1 per cent.*

---

* Natural carbon also contains an infinitesimal amount of carbon-14, used in determining the age of antiquities. It has eight neutrons, and is radioactive.

The specimen of carbon bombarded by the neutrons would thus be made to contain more atoms of carbon-13 than occur in natural carbon.

On the other hand, if we bombarded sodium, which has only one stable isotope and in which all the atoms contain 11 electrons, 11 protons and 12 neutrons, we would produce an unstable form of sodium with 13 neutrons. One neutron is surplus and cannot be held in the nucleus; it can change from the unstable condition by acquiring a positive charge which turns it into an extra proton. Since the neutron is an electrically neutral particle, it achieves this transformation by giving out a negative charge (i.e. an electron). This brings the atom to a condition where its nucleus now contains 12 protons (an increase of one) and 12 neutrons (same as before), and therefore the atom is no longer one of sodium, but has been converted to the next element, magnesium, which has 12 electrons.

Realizing that neutron bombardment could convert an element into one with a higher atomic number, the Italian physicist Enrico Fermi had the idea of creating new elements by bombarding uranium, which, with 92 protons, was the largest known atom. The metal is composed principally of uranium-238, but less than 1 per cent of its mass derives from the isotope uranium-235. Following the principle described above for converting sodium into magnesium, Fermi succeeded in proving that he had converted part of the uranium, first into neptunium, with 93 protons, and this in turn into plutonium, with 94 protons; both these are synthetic elements.

In 1938 the German physicists Otto Hahn and Fritz Strassmann were studying the bombardment of uranium by neutrons, and repeated Fermi's experiments. At first they thought they had produced only 'transuranic elements' (ones heavier than uranium), like plutonium, but on repeating the experiments they discovered that other, lighter, elements had also been formed, including the metal barium and an inert gas, probably krypton, both in such minute amounts that their existence could barely be identified. An Austrian scientist, Lise Meitner, and her nephew Otto Frisch, interpreted these unexpected results as indicating that the nucleus of the uranium-235 isotope had been split into several fragments with atomic weights ranging from about one-third to two-thirds that of uranium, coupled with the production of several secondary neutrons. In contrast with this, Fermi's experiments had only revealed that the atoms of uranium-238, comprising the major part of the metal, had been changed into slightly larger atoms, neptunium and plutonium. Hahn and Strassmann's momentous findings, as interpreted by Lise Meitner, showed that they had also achieved 'fission' of the 235 isotope, which behaved in a different manner from the 238 isotope.

If the formation of the fragments were the only result of fission of the uranium atom, the process would be of no greater value in the provision of large-scale energy than were Rutherford's early experiments. But the fission of uranium-235 also releases two or three neutrons. Thus we reach the position envisaged on page 252, in which the splitting of one atom provides two or more projectiles well suited to split more atoms, initiating a chain reaction capable of continuous self-activation, with the release of energy. The weight of the nucleus of uranium-235 is about one thousandth part heavier than the sum of the weights of the fission fragments. Einstein's equation states that $E = mc^2$, where E is a measure of the amount of energy liberated when a mass m is annihilated; c is the velocity of light. If the whole matter of only one gram of a substance could be destroyed, energy equivalent to over two hundred million kilowatt-hours would be produced.

Only a few days before Hitler invaded Holland, Belgium and France, this discovery was reported in the newspapers; immediately afterwards, for reasons which are now well apparent, further references to the fission of uranium-235 were put under the seal of security until after Hiroshima, over five years later.

URANIUM ENRICHMENT

The amount of uranium-235 contained in the small sample bombarded by Hahn and Strassmann was minute, the neutrons were fast-moving and most of them escaped, so a chain reaction could not take place. During the following decade efforts of scientists were concentrated on the technique of producing large amounts of ultra-high purity uranium-235 so that methods of violent destruction could be achieved.

When attention was turned to the harnessing of nuclear energy for peaceful purposes, it was realized that uranium only slightly enriched with its 235 isotope could be used. It soon became obvious that metallic uranium has some physical properties which make it difficult to contain in a reactor, and enriched uranium oxide was found to be more suitable. First it was necessary to discover ways of enriching the uranium more economically than the fabulously expensive gaseous diffusion method used in the 1940s.

The chemical properties of the isotopes of an element are identical, so separation must rely on the slightly different atomic weights of the isotopes. One method is to make the compound uranium hexafluoride, which when warmed becomes gaseous. The molecules of the gas containing uranium-235 are a little lighter than the major part containing uranium-238 and, in the gaseous form, move about or diffuse at a slightly greater rate than the heavier ones, the difference being about 1 per cent.

The gaseous uranium hexafluoride is made to diffuse through very fine pores in a series of multiple diaphragms. The faster-moving molecules pass through more readily than the heavier ones, so, as the gas diffuses through each diaphragm, the proportion of the lighter molecules is increased slightly. The process, repeated over and over again, gradually increases the proportion of the 235 isotope in the gas.

Uranium hexafluoride is highly corrosive and will destroy many of the materials of which chemical apparatus is made. The compound is liable to decomposition, particularly if moisture is present, and unless stringent precautions are taken it will change to a solid material deposited on the diaphragm membranes, evolving the poisonous fluorine and hydrofluoric acid. It is an abominable substance, nicknamed 'Hex' – which is also an evil and destructive spirit.

Until the 1960s, gaseous diffusion remained the principal method of enriching uranium. However, as far back as 1940 the 235 isotope had been separated experimentally by centrifuge, but at that time the technology of high speed rotating machinery was not equal to the task. Improvements in centrifuges led to the establishment of enrichment plants in which the amount of electricity used, though still enormous, was less than one-tenth that required by the wartime methods of gaseous diffusion. In 1970, Britain, The Netherlands and West Germany began to expand the gas centrifuge process. One pilot plant was built at Capenhurst in the UK, while the Dutch and German plants were built at Almemo and Gronau. In the USA a centrifuge plant was closed and there was a return to enriching by diffusion, but now there is promising US research on laser enrichment. This requires the uranium to be vaporized; a laser ionizes the 235 isotope but not the 238 one. The vapour is collected and condensed, and yields an enrichment ready to make uranium dioxide powder.

In the centrifuge process uranium hexafluoride is heated to 80°C to convert it into gas, which is fed into a series of vacuum tanks, each containing a rotor about 1 m long. When the rotors are spun rapidly the heavier molecules, based on the 238 isotope, move towards the circumference more readily than the lighter molecules containing uranium-235. In modern counter-current centrifuges, 'scoops' inside the drum make it possible to feed the gas enriched in 235 and the depleted gas to separate exits at the top. Each rotor enriches one tenth of a gram of Hex per second; the process is repeated over and over again in cascades of several thousand rotors.

The peripheral speed of each rotor needs to be at least 400 m per second. The basic object in designing a centrifuge is to provide the fastest and longest rotor possible, with maximum life and minimum cost. The rotors must be of very strong material. Stainless steel and some aluminium alloys will withstand 400 m per second, titanium alloys up to 440 m per second and maraging steels (page 190) up to 525 m per second. For still higher speeds a fibreglass/polyester composite has been developed in Europe. Thousands of miles of piping and at least 100 000 welds are required in a centrifuge plant, and in the coming years several million centrifuges will be required.

CONTROLLED FISSION IN NUCLEAR POWER STATIONS

It is worth reminding ourselves that in the atomic bomb, which was made of over 90 per cent uranium-235, fast neutrons initiated the fission. So long as the mass of uranium-235 was less than critical, the fast neutrons would escape before having a chance to cause fission. When the critical mass was exceeded the fast neutrons became so thoroughly trapped that the violent chain reaction began immediately. Fast neutrons travel at speeds of many thousands of kilometres per second. In the controlled conditions of the nuclear reactor slow neutrons, having speeds of only a few kilometres per second, are harnessed. Such neutrons are commonly called 'thermal neutrons'. Some of these hit and split uranium-235 atoms, thus bringing about a chain reaction. However, unlike the condition in the atomic bomb, this is a controlled chain reaction, regulated by materials which can absorb neutrons.

The 235 isotope which accounts for 0.7 per cent of natural uranium can be split by slow neutrons which divide the isotope into fragments, creating energy and releasing fast-moving neutrons. These must be slowed down by what is called a moderator. Ordinary water is an effective moderator, but it also captures some of the neutrons, so before it can be used in a nuclear reactor the metal must be enriched to about 3 per cent of uranium-235, using methods described above. Graphite and heavy water capture far fewer neutrons, so if either of them is used as a moderator, controlled fission can take place, using uranium metal or its oxide, if necessary unenriched.

Since the first nuclear power plants were commissioned there have been many new designs, some experimental, some for submarines and some for industrial power stations. Each has its own specific problems of engineering, metallurgy and control. First we look at the early gas-cooled reactors, then the most popular system, the pressurized water reactor, and finally the fast-breeder reactor which, though little used at present, may eventually offer great opportunities for efficient operation.

EARLY GAS-COOLED REACTORS

Figure 45 is a schematic drawing which illustrates the principles of the first gas-cooled reactor power station. Four reactors are housed in cylindrical steel pressure vessels, surrounded by a concrete shield. The uranium is sealed in metal 'cans' to contain the waste products of fission. Each can is finned to assist the transfer of heated carbon dioxide from the reactors to the heat exchangers. The cans for containing the fuel elements are made of a magnesium alloy (which is discussed later).

The speed of the neutrons is controlled to prevent excessive heat from being generated. This is done by selecting a material which can entrap neutrons; the control rods made of it can be adjusted in position to maintain a steady 'neutron cross-section' in the reactor. During normal operation the control rods are moved within the reactor core by electric motors; this movement is finely controlled to fractions of a millimetre. The control rods are suspended electromagnetically so that, in the event of electrical failure, they would drop fully home, their full length of nearly ten metres then being entirely in the reactor core; this would immediately stop the chain reaction. In the reactors at Calder Hall there are 48 control rods, made of a steel containing boron. The heat of the reactor is taken away by forcing carbon dioxide under pressure through the system. The gas enters the reactor at 140°C and is heated to about 340°C.

Other stations, usually called 'Magnox' after the magnesium alloy of which the cans were made, were constructed in Britain. After the first Magnox reactor was getting into its stride, a new concept known as the advanced gas-cooled reactor (AGR) was being constructed. This used uranium dioxide as fuel, the uranium being slightly enriched to about 2.5 per cent of the 235 isotope, giving a longer life to the fuel and allowing the carbon dioxide temperature to be 660°C, about 300°C higher than in the Magnox reactors.

As the magnesium alloy cans are not suitable for the greater heat, the uranium dioxide pellets are inserted in stainless steel cans. *Figure 46* is a diagram of one element. Each element contains 36 stainless steel tubes arranged in three rings, of 18, 12 and 6 tubes. The assembly is surrounded by a graphite sleeve. The AGR contains about three hundred channels, each holding eight fuel elements connected together lengthwise. The fuel containers, guide tubes, support grids and centre braces are of stainless

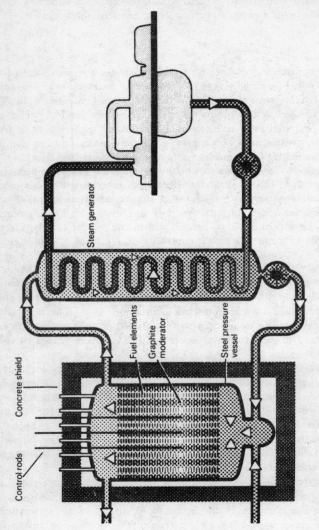

Fig. 45. Basic gas-cooled reactor (Magnox) (*Courtesy British Nuclear Fuels plc*)

Control rods

Concrete shield

Fuel elements

Graphite moderator

Steel pressure vessel

Steam generator

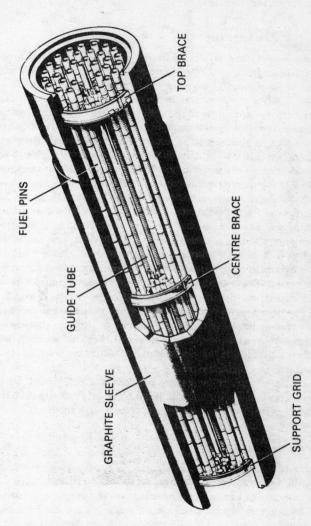

TOP BRACE

FUEL PINS

GUIDE TUBE

CENTRE BRACE

GRAPHITE SLEEVE

SUPPORT GRID

Fig. 46. Advanced gas-cooled reactor fuel element (*Courtesy British Nuclear Fuels plc*)

steel containing an added stabilizer element, titanium for the centre brace and niobium for the others.

PRESSURIZED WATER REACTORS

A type of compact reactor developed in the USA for submarine propulsion led to the emergence of the pressurized water reactor (PWR), which has become the world's most popular system. At present there are over two hundred PWRs in operation in more than twenty countries. In France alone there are over fifty, providing three-quarters of that country's electricity. France is continuing to build PWRs, and plans to build a new series in about 2010 to replace the old stations. Britain has been slow to adopt this system, but one reactor at Sizewell is due to open in 1994. In addition to the many power stations using this system, there are over 350 PWR-powered naval vessels in service.

A typical PWR structure is shown in *Fig. 47*. It consists of a large steel pressure vessel which contains the reactor core and control rods. The remaining space is occupied by ordinary water, held at a pressure of 150 atmospheres so that, although it reaches a temperature of over 300°C, it does not boil. The fuel pins arranged in clusters are of a zirconium alloy which has great resistance to corrosion when surrounded by very hot water. The 'pins' are only 1 cm in diameter and 3.8 m long, and contain cylindrical pellets of uranium dioxide, enriched to 3.2 per cent of the 235 isotope; 264 of these pins are assembled in a square array. The nuclear reactor heats the water, which passes through thousands of tubes, immersed in water in a second vessel, held at a much lower pressure, forming steam which is piped to a turbine coupled to the electric generator.

One major difference in technology between the PWR and the gas-cooled reactors is that there is no graphite moderator, since the circulating water acts as both moderator and means of transferring the heat.

FAST BREEDER REACTORS

A different concept in which no moderator is required is illustrated in *Fig. 48*. In the fast reactor ('fast' refers to the speed of the neutrons), a mixture of oxides of uranium and plutonium is clad in stainless steel and made into compact hexagonal fuel element assemblies. By including a blanket of uranium waste product from enrichment plants, that material is converted by neutron capture, first into neptunium and then into plutonium-239, which is a fissionable isotope; the process is known as

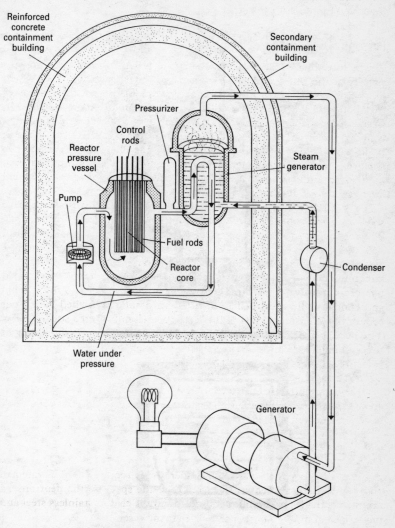

Fig. 47. **Pressurized water reactor**
(*Courtesy British Nuclear Fuels plc*)

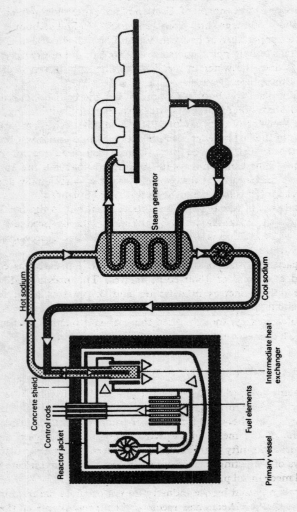

Fig. 48. Sodium–cooled fast reactor (*Courtesy British Nuclear Fuels plc*)

Hot sodium

Cool sodium

Concrete shield

Control rods

Reactor jacket

Intermediate heat exchanger

Fuel elements

Primary vessel

Steam generator

breeding. The considerable amount of heat evolved is transferred by molten sodium which, having a high thermal conductivity, is an excellent material for conveying heat from the reactor core.

At Dounreay in the north of Scotland, an experimental plant was producing about 60 megawatts from 1959 to 1977. The sodium coolant operated at a temperature of between 500 and 600°C, about 300°C lower than the metal's boiling point, so its use as a coolant did not require pressurization, as does the water in the PWRs. The sodium was contained in an argon atmosphere.

Later a pilot plant was built at Dounreay, but the British Government was advised that other energy sources were not likely to be come depleted or over-costly until about 2030, so it was decided to suspend development of the fast breeder reactors. France built a pilot fast breeder reactor in the 1970s, followed by a commercial-sized reactor called Super-Phenix. This came into useful production in 1989, but in 1992 France halted further use of fast breeder reactors, and it now appears that only Japan intends to continue with this type of plant.

The capital cost of fast breeder reactors is high, but in the far future they may have the advantage that, by the use of the by-products from pressurized water reactors as fuel for the fast reactors, scarce uranium reserves may be made to last longer.

The comparison of sizes of reactors is significant. The early gas-cooled reactors had a volume of 1300 cubic metres. The pressurized water reactors are only 3 metres in diameter and 3.7 metres high, which amounts to about 25 cubic metres. The reactor in the fast breeder is only about 4.5 cubic metres, though when the core and blanket of uranium waste are taken into account, they occupy about 9 cubic metres.

THE ROLE OF THE METALLURGIST IN NUCLEAR ENGINEERING

All the systems for producing power from nuclear fission pose new problems for the metallurgist – the fissile materials, the control rods, the canning alloys and the extraordinary corrosion conditions inside the reactor. Some of the metals now used quite extensively were merely laboratory curiosities fifty years ago, and it has been necessary to discover the best ways of extracting them from their ores and to investigate their physical and mechanical properties.

Some metals required by the industry do not occur in nature, but are made in laboratories. When a new reactor is being commissioned there is usually insufficient 'neutron density' to cause it to start up: a neutron emitter is required to give it a 'kick start'. At first the synthetic radioactive element polonium was used for this purpose, but now an isotope of californium is preferred.

Metallurgically, uranium has a number of peculiarities. It oxidizes rapidly in air at 200°C and above; its atomic lattice structure at room temperature is complex, and each crystal of the metal expands non-uniformly when heated, leading to internal stresses when the metal undergoes heating and cooling. Another distortion problem connected with the properties of uranium is its growth when exposed to radiation: one direction in the uranium crystal gets longer, while another contracts. It is therefore rather unsuitable for use in the metallic form.

The fuel in most modern reactors is uranium dioxide, made by sintering it in powder form at 1650°C in an inert or slightly reducing atmosphere. The finished product has about 95 per cent of the possible theoretical density. During the early stages in a reactor, further sintering occurs and the fuel pellets shrink a little. Although this shrinkage has a negligible effect on the diameter of the pellets, the change in total length may add up to a centimetre. A problem would arise if part of the stack of 64 pellets got stuck in a cladding tube, leaving a gap between the pellets. This is overcome by machining an 'anti-stacking groove' in some of the uranium dioxide pellets. During manufacturing the fuel pins are pressurized and indented into these grooves, locking the pellets in position.

Great strides have been made in the quality of stainless steel to meet the demands of the nuclear industry. This has also led to new methods of making thin stainless steel cladding tubes, and there have been important developments in welding and inspection techniques.

CANNING MATERIALS

Magnesium was a suitable can material for the first designs of gas-cooled reactor. A magnesium alloy known as Magnox, containing about 0.8 per cent aluminium and 0.01 per cent beryllium, was used to sheathe the uranium in those reactors. The small alloying additions ensured that a tenacious and protective oxide skin was formed. As processes which evolve more heat were developed, different canning materials were required and pellets of uranium oxide replaced the pure metal. The containers became more slender, and are now generally called 'pins'.

Stainless steel has the advantage of resisting heat and not warping, though to get optimum results the material has to be only a third of a millimetre in wall-section thickness. The advanced gas-cooled reactors in Britain use a steel with 20 per cent chromium, 25 per cent nickel and

1 per cent niobium. *Plate 29* shows a selection of fuel elements used in various nuclear power stations.

Zirconium alloy pins feature in the pressurized water reactor and in several other reactor designs. Two alloys are specified, with similar compositions. Both are based on zirconium with 1.2–1.7 per cent tin; one also contains 0.07–0.20 per cent iron, 0.05–0.15 per cent chromium and 0.03–0.08 per cent nickel; the other contains 0.18–0.24 per cent iron, 0.07–0.13 per cent chromium and no nickel.

MISCELLANEOUS PROBLEMS

The advanced gas-cooled reactors have two types of control rods, either boron steel inserts in stainless steel sheaths or simply rods of stainless steel. Recently a control rod alloy consisting of 80 per cent silver, 15 per cent indium and 5 per cent cadmium has been developed for pressurized water cooled reactors, where optimum resistance to corrosion by the coolant is necessary. For specialized compact reactors other materials, usually exceedingly costly ones, are chosen for their great ability to absorb neutrons. Since hafnium became more available, it is being specified for some compact reactors. By far the best absorber of neutrons is the rare earth metal gadolinium (page 216), but supplies are so scarce that its use is limited to special reactors which require control rods of extraordinarily high efficiency.

Pressurized water reactors operate at such high pressures that steel vessels are essential often with a wall thickness of over 200 mm; the welding of this plate and its heat treatment are precise and important operations. Special attention must be given to 'compatability', which involves similar effects to sacrificial corrosion (page 227). The metals in the nuclear power unit must not deteriorate through contact with other materials. For example, stainless steel components in fast reactors had to 'sit' in molten sodium and be exposed to fast neutrons, all in addition to the thermal stresses which must be endured.

A notable feature of nuclear metallurgy has been in the field of what is known as post-irradiation examination. Materials can be removed from the reactor, sectioned, metallurgically examined or subjected to tensile tests without ever handling the metals or tools. Specimens are treated in vacuum furnaces, welds are made and X-rayed and photographs are prepared, all totally by remote control. Automatic machines have been developed which can remove fuel elements from reactors at various stages by remote control, so that they can be measured and any unfavourable trends detected at the earliest possible stage.

27

THE FUTURE OF METALS

Niels Bohr once said, 'Prediction is very difficult, especially when it is about the future.' Somerset Maugham was pessimistic: 'It's bad enough to know the past, it would be intolerable to know the future.' We are more optimistic: we are convinced that the field of metallurgy will continue to be well cultivated, and fruitful. Large tonnages of metals will continue to be required for vehicles for land, sea and air, and also for an increasing variety of domestic appliances, and for all the girders, tubes and wires in factories, power stations and sports stadia. More metals will be needed for the agricultural machinery of developing countries, for untold quantities of the steel mesh and rods that strengthen reinforced concrete, and for the beautiful shapes that make bridges.

It has been forecast that by the year 2050 the world's population will be twice what it is in the 1990s. We can only guess how much this will affect the production of materials, because so much depends on the future development of China and the Third World countries. At present there are only about 30 000 cars in China (though many more ancient trucks). If the population of that vast country comes to own automobiles, the extra production of steel and aluminium will be enormous.

In the Western world the consumption of iron and steel may decline, not because of a lower standard of living but because more efficient design and a greater use of alloy steels will lead to lighter components. The trend to reduced weight has been continuing since the Industrial Revolution. In 1810, when boilers were made of cast iron, the ratio of weight (in tons) to power (in horsepower*) for a steam engine was 1000:1; by 1900 it was 100:1. When electric locomotives were introduced the ratio became 25:1 and by 1990 it was only 13:1. During those years the weight reduction achieved by improved design and higher grade

---

* 1 horsepower equals 745.7 watts.

materials was more than counterbalanced by the vast increase in the use of ferrous metals in cars, bridges, buildings and ships, so the total consumption of iron and steel increased. However, it is not likely that many new immense steel-using industries will develop in the future.

The probable fall in the consumption of iron and steel in the West can be illustrated by the greater use of alloy steels in the automobile industry. In 1975 about 5 per cent of the steel in the average car was high strength or stainless; by 1985 the proportion was 14 per cent, and is expected to rise to 20 per cent by the year 2000. Such changes are accompanied by a reduction in the tonnage of steel, because each kilogram of alloy steel replaces 1.3 kg of ordinary steel.

Most manufactured articles in other metals are tending to become lighter. For example, the weight of each aluminium can decreased by about 25 per cent from 1981 to 1991, though the total weight of aluminium used by that industry increased massively because of the greater number of cans produced. The automobile industry is requiring more aluminium per car each year. In the early 1990s, Audi was advertising an 'all-aluminium' car, which may indicate the shape of things to come. Consequently, although each component may be lighter, the total amount of aluminium will certainly increase.

Iron and aluminium will still be produced in the greatest tonnages, but we wonder whether they will continue to be the 'most important' materials. A world without the silicon chip is now inconceivable. Sooner or later, when nuclear energy is widely accepted and there are no more Chernobyls, uranium may become one of the most important metals. Later on, when fusion replaces fission, lithium may become essential.

SOURCES OF METALS

Much has been written about forthcoming world shortages of copper, lead, zinc, silver, cadmium, cobalt, antimony and mercury. They will become increasingly expensive, and their uses may be limited to specific purposes for which they cannot be replaced. We must add that, fortunately, deposits of iron, aluminium and magnesium ores will last for a very long time.

The sea floor offers an exciting possibility as a source of metals. Mineral nodules like large pebbles have been found on the sea floor in concentrations of 15 000 tonnes per square kilometre at depths of 3000 m. The origin of these nodules is obscure. However, wherever two tectonic plates of the Earth's mantle are drifting slowly apart, and lava from the interior pushes out into the sea, compounds of manganese, iron, nickel and other elements are leached from the lava and it is thought that these form the sea floor nodules. Preparations are being made to dredge areas between southern California, Panama and Hawaii, the object being to obtain nickel, copper and cobalt. But does the sea floor belong to any one country, or is it the common heritage of all of us? Whoever establishes the right to 'mine' it will be faced with costly and difficult dredging operations.

Looking into the far future, it is possible that metals becoming scarce on Earth may be obtained from extraterrestrial sources. One of our closest visitors, the small asteroid 1986 DA, is made mostly of iron with 8 per cent nickel, but it also contains ten parts per million of the precious metals gold, platinum and palladium, representing about 100 000 tonnes of these valuable materials. (Strangely enough, compared with the mass of the asteroid, this is the same proportion as that of gold in the deep mines of South Africa.)

In the first edition of this book there was a chapter discussing strategic metals in wartime. *Figure 49* was drawn to illustrate the tug-of-the-war between the Allies on one side, and Germany, Italy and Japan on the other. Even if wars do not occur on a global scale, local conflicts seem inevitable and political battles often involve strategic materials. For some years the USA would not purchase tungsten from China, which holds about half the world's reserves. Eventually relations were improved and tungsten was bought direct from China, instead of through intermediaries.

Dependence on supplies of metals from countries with political problems can be understood by referring to *Table 3*, on page 12. Chromium is obtained from South Africa, Albania and Zimbabwe. Cuba has large reserves of nickel ores. Major supplies of vanadium come from South Africa and Namibia. Often the total economy of an otherwise underdeveloped country depends on its ability to excavate and sell supplies of an important metal ore to the rest of the world. The conflicts endured by Zaire and Zambia in the late 1970s showed how such economies can be wrecked, resulting in worldwide shortages and large increases in, for example, the price of cobalt.

NEW ALLOYS

Metallurgists have tended to concentrate their researches on the major industrial metals: iron, aluminium, copper, magnesium, nickel, titanium, tin, lead and zinc, but future progress will come from the study of lesser-known metals. Out of the Dramatis Personae at the beginning of this

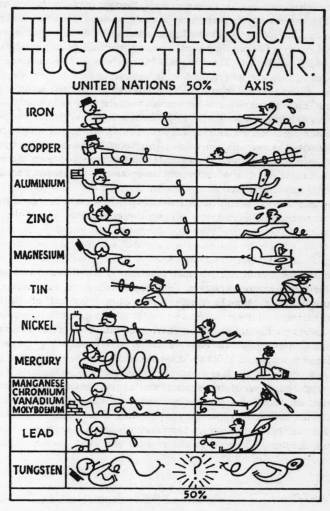

Fig. 49. The metallurgical tug-of-the-war, Autumn 1943 (*Reproduced from the first wartime edition of this book, courtesy Joanna Coudrill*)

book, 68 elements have been proved to be valuable in alloying (the others were radioactive, very rare or synthetic). In principle, these 68 could form 2278 binary alloys, perhaps 1500 of which would have possibilities. If ternary alloys were considered, there would be 50 116 combinations, of which, say, 30 000 might be viable; quaternary alloys might well offer over 100 000 of interest. The list could be enlarged by combinations with phosphorus, nitrogen and hydrogen. In each alloy system there would be a range of composition from nearly zero to nearly 100 per cent for each element. Thus to investigate the near-infinite number of possible alloys, metallurgical research would have a full programme.

So many new and interesting alloys are being developed that it would be impossible to do justice to even a small fraction of them, so we have chosen four groups which so far in this book have not been discussed in detail. The first is of universal importance; the second contains some new alloys required in supersonic aircraft; the third is admittedly of only minor importance, but we found it rather fascinating; the fourth introduces 'Buckyballs', some of which contain metals.

Magnetic alloys

There are about a hundred small magnets in the average household. Without them we would have no television, no deep freezers, no microwave ovens, no automobiles, no calculators, no electric clocks, and no digital watches. All these and many other devices depend on the operation of highly magnetic materials, and there is a constant impetus to discover new and better ones. An early magnetic alloy containing cobalt was mentioned on page 204. Then the rare earth metals entered the field, with samarium–cobalt and boron–iron–neodymium. These alloys can develop magnetic properties more powerful than any previously known materials. In addition to the endless search for new combinations, there is much development work to make the present alloys more efficient, to improve their behaviour at high temperatures and to improve methods of manufacture.

Aluminium–lithium alloys

Lithium is the lightest of all metals, but it is highly chemically reactive. Experimental work on aluminium–lithium alloys was begun in the 1950s, but it was not until the 1980s that production problems were overcome and bulk output achieved. Currently the most favoured alloy contains 2.5 per cent lithium, 2.4 per cent copper and 0.75 per cent magnesium, with

0.12 per cent zirconium for grain size control. Such an alloy is 10 per cent lighter and 10 per cent stiffer than the conventional high-strength aluminium alloys, thus enabling designers to make a weight saving of up to 15 per cent in an aircraft component. At present, the aluminium–lithium alloys are several times more costly than other aluminium alloys, but their advantages justify their use in space satellites and the airframes of supersonic aircraft.

Shape-memory alloys

In the late 1970s the shape-memory alloys were discovered. Components made of such alloys which are plastically deformed at one temperature and then warmed will change in shape but return to their original form when the temperature is lowered again. Later it was discovered that by suitable processing some alloys could be 'trained' to exhibit memory, with complete reversibility on heating and cooling.

An early use was for a greenhouse window-opener, with a coil spring of a copper–zinc–aluminium alloy which opened and closed the window in response to changes in ambient temperature. This alloy also was used for coffee-makers and thermal sensor–actuators to control the flow of hot water in solar heating systems. The shape-memory effect occurs in several other alloy systems, such as nickel–titanium, copper–aluminium–nickel and gold–cadmium. Future developments will probably also involve an iron–nickel–cobalt–titanium alloy. In Japan, a nickel–titanium alloy is used in the manufacture of underwired brassières so that they remain comfortable over a range of temperatures, and one company planned to use a shape-memory alloy in thin wire form to give trousers a permanent crease.

Fullerenes, known as 'Buckyballs'

The scientific world has long known that carbon exists in the radically different forms of graphite (the smoothest) and diamond (the hardest). That such a well-known and important element, the basic component of all life, should also exist in a third form came as a great and exciting surprise. In May 1990, scientists discovered a sphere-like molecular structure of carbon atoms, distributed in a regular pattern, like the stitching of a soccer ball. The best-known structure contains 60 carbon atoms, but some have been produced with 70 atoms (distributed like the outside of a rugby ball), and several other forms with more carbon atoms.

These molecular structures closely resemble the geodesic dome, a structural space-frame developed by American engineer R. Buckminster Fuller, so the sphere-like molecules were named 'fullerenes' or, more colloquially, 'Buckyballs'. The C_{60} molecule is the most symmetrical known, and spins at over 100 million revolutions per second. The year after their discovery, worldwide research began to explore the possibilities of the new materials. The carbon molecule C_{60} could be doped with potassium, forming the molecule K_3C_{60} which has high electrical conductivity. If a potassium atom is placed at the centre of the Buckyball, the combination becomes superconductive (though so far it does not operate at temperatures as high as those which will be described on page 294). Superconductive experiments are continuing with such combinations as potassium–rubidium–thallium-doped Buckyballs. The molecule doped with fluorine, forming $C_{60}F_{60}$ promises to become one of the best lubricants ever discovered. Some Buckyballs are as soft as graphite, but it has been claimed that, when compressed to 70 per cent of their initial volume, they become harder than diamond.

The next few years will show whether fullerenes, combined with metals or non-metals, have exploitable properties. Manufacturing costs are not expected to be high. The fullerenes can be made by setting up an electric arc between two graphite electrodes in an atmosphere of helium, so the cost should be comparable to that of making aluminium.

NEW PROCESSES

We saw in Chapter 8 that, whether it is a needle or a railway line that is being made, manufacturing processes are ever changing and ever improving. Here we mention two technologies which have become established but which offer promise of even greater developments in the future.

Lasers

A laser (light amplification by stimulated emission of radiation) consists of three components. The material which will be made to 'lase' can be a solid, a liquid or a gas. Ruby was the first laser material, but industrially a neodymium-doped yttrium–aluminium garnet (known as Nd–YAG) is now commonly used, as is gallium arsenide. A liquid such as a dye dissolved in ethylene glycol is yet another source. For very powerful laser beams, carbon dioxide is pre-eminent. The second component of a laser is the source of excitation, which can be electric power or light. The third is an amplifying arrangement which normally consists of two parallel mirrors positioned so that the light beam can 'bounce' between them, becoming amplified in strength and made coherent in phase. The amplified beam is highly directional and of one wavelength.

Nowadays, Christmas displays in the high street, rock concerts and spectacular musicals are embellished by laser displays, but lasers have become important in many other fields, from manufacturing to surgery and cartography. There is one in every compact disc player. The power of those used in industry ranges from less than a watt to over 25 kW.

Medium power Nd–YAG lasers of up to 500 W are used for drilling, scribing and engraving metal articles. Carbon dioxide lasers from 500 to 1500 W are suitable for profile-cutting flat sheet metal or for three-dimensional trimming of automobile body panels. Very high power carbon dioxide lasers from 1 to 25 kW are used for high speed welding and for heat treatment of such parts as automobile camshafts. In the manufacture of gas turbines, lasers are used to drill small holes in such parts as nozzle guide vanes and combustion chambers. Rolls-Royce found that in such applications a cost saving of about 30 per cent could be achieved. Because of its great accuracy and absence of burrs in the product, the laser tool eliminates such operations as milling, drilling and de-burring. Wherever automation is involved, the laser can be computer-programmed.

The use of lasers for welding has begun, but there are still problems to be overcome. Carbon dioxide laser welding permits high speed of operation, very low distortion of the material and the possibility of welding several components at the same time. Very close abutment of the parts to be joined is essential, requiring special tooling; furthermore the equipment is costly. Thus laser welding is being used principally where there is automation and mass-production, as in the manufacture of automobiles.

Lasers can be used to separate isotopes. A beam, tuned specifically to the frequency of the atom of one of the element's isotopes, can enable some proportion of that isotope to be separated from the mixture, leaving the atoms of the other isotopes unaffected.

Amorphous metals

Rapid solidification (RS) technology began in the late 1960s as a rather fantastic idea, but it has now become an established metallurgical process with promise of further development. A molten alloy (typically iron–boron–silicon) is cooled from over 1000°C to ambient temperature in just one millisecond. In one method a stream of the alloy is forced through a nozzle from the feeder head and falls onto the chilled surface of a rapidly spinning copper wheel, from which it is immediately rolled into thin strip and wound onto a reel ready for delivery to the user.

In this way the alloy cools so rapidly that the atoms do not have time to form a crystalline pattern. The amorphous structure has no grain boundaries, and is extremely hard and less vulnerable to corrosion than conventional ferrous materials. Furthermore, the absence of grain boundaries gives amorphous metals unusual magnetic properties.

Early RS experiments were done with a silicon–gold alloy, but since then many materials have been evaluated. Most are based on iron with additions of silicon and boron, for example 92 per cent iron, 5 per cent silicon and 3 per cent boron. The two non-metals are added as ferro-alloys, and it is important that impurities such as aluminium, manganese, calcium and titanium, which can be absorbed in the production of the ferro-alloys, are kept to under 0.01 per cent.

The magnetic properties of amorphous metals are valuable in the manufacture of transformers for power transmission, where typically 100 kg of core is required for each transformer. With the US annual production of half a million transformers, the prospects for the use of amorphous metal cores are interesting. These RS alloys, which compete with silicon steel in transformers and electric motors, are also candidate materials for cassette recorders, computers and aircraft.

CONSERVATION AND RECYCLING

Some metals have important uses which make their conservation difficult. Tungsten is required for the hundreds of millions of tool-tips and drills required by factories all over the world. During the life of those tools the metal is gradually dispersed and is therefore impossible to reclaim. Tungsten elements are a feature of electric light bulbs; one wonders what could be done to separate this metal from the glass bulbs and brass holders. Zinc is used for the sacrificial protection of offshore oil rigs and, though it performs an excellent and essential service, it is indeed sacrificed.

It takes several times more energy to mine, transport and smelt iron ores than to collect and remelt ferrous scrap. Aluminium, which requires a vast amount of energy in its production, cries out for conservation because the energy cost of reclaiming aluminium scrap is only one-twentieth the cost of making the pure metal. The thousands of aluminium collection centres in many parts of the world show that the problem is being taken seriously. Many good causes and children's pocket money have benefited from the small payments offered for the return of aluminium cans.

In the past, the source of scrap metals was the rag-and-bone man's collection of old saucepans, bedsteads, broken-down lawn mowers and cheap tin trays. Now the collection, sorting, shredding and reclamation of materials is a large and important industry. A great deal of valuable material comes from automobiles. For many years they have been mass produced on flowline systems, but during the final decade of this century the wheel is coming full circle: old cars will move along automated *dis*assembly lines. There are about 140 million cars on the roads of Europe, and about 12 million of these, weighing a total of 12 million tonnes, reach the end of their life each year. They contain about 75 per cent metal (mostly ferrous), but the remainder, totalling about 3 million tonnes in Europe alone, has to be consigned to landfill.

Fortunately, the European Community has designated car recycling a priority, to reduce the amount of natural resources that is wasted and to minimize the amount of pollution. The German Government has led the demands for manufacturers to take responsibility for their products 'from the cradle to the grave'. Now designers are planning their products for easy dismantling and identification of components to facilitate recycling. To paraphrase the remark we made at the end of the chapter on corrosion, 'conservation begins on the drawing-board'. Special tools and techniques are being developed to dismantle cars rapidly, and this will introduce the problem that, in contrast to the 'one model' assembly line, the disassembly line may have to deal with the individual construction, and therefore destruction, of a range of vehicles.

First, petrol, oil and brake fluid are removed. Heavy items such as batteries, cylinder blocks, gear boxes, alternators, starters and electronic units are sent to specialist companies. In the next five to ten years catalytic converters will be coming onto the disassembly lines; they will contain platinum, palladium and rhodium.

A fragmentizer smashes the material into fist-sized pieces. A magnet separates most of the ferrous metal and then a cyclone, like a huge vacuum cleaner, takes off the light non-metallics, leaving the heavier non-magnetic material, which is a mixture of non-ferrous metals plus rubber, glass and large plastic parts, together with dirt and stones. This is dealt with by what is known as the sink-and-float process – also called the heavy media plant. The material is flooded with water to take away pieces of dirt and textile material. Then the scrap passes to tanks containing suspensions of magnetite and ferro-silicon in water, which can be adjusted in composition to produce a specific gravity between 1.25 and 3.8. Any remaining flock and upholstery float on the liquid, while the metals sink. In a further stage, with the higher density fluid, aluminium floats, while zinc, copper and brass sink, to be removed later.

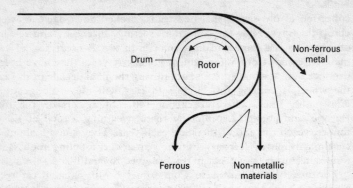

Fig. 50. The Eriez Cotswold Separator
(*Courtesy Cotswold Research Ltd and The Bird Group*)

Although this process provides a useful separation of heavy and light metals, there is still the problem of aluminium being entangled with other materials and with small pieces of ferrous metal that were not removed magnetically. Furthermore, some non-metallic material has not been segregated.

The first successful way of removing the light fraction from the heavy media plant was to use a separator based on a linear induction motor which generated an eddy current at right angles to the lines of the conveyor. This has been superseded by the Eriez Cotswold Separator. A high speed magnetic rotor in the conveyor drum causes any remaining ferrous metal to fall in one direction, aluminium falls away from the ferrous metal, and non-metallics fall between the two – as shown in *Fig. 50.* This process enables the separation of various metals, including small pieces, which was not previously possible. The reclaimed aluminium is of sufficient purity for making into the high performance alloys required by the automobile industry.

Some of the problems of separating the materials from which a car is built can be appreciated by the two items shown in *Fig. 51:* the small door mirror, containing nine different materials, and the large bumper assembly, containing only four. It must be admitted that a 'headache' for the 1990s will be the separation and reclaiming of the hundreds of different plastics. They do not have the magnetic properties or the substantial differences in specific gravity which make it possible to separate individual metals. As a first step, each plastic item on an

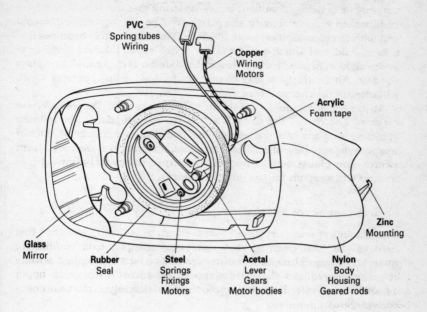

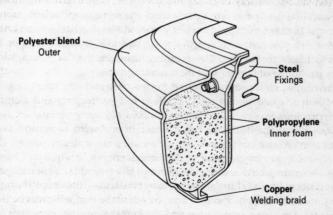

Fig. 51. A small component like a door mirror contains nine different materials. A bumper, although large, contains only four materials and is a prime candidate for recycling.
(*Courtesy the Bird Group*)

automobile is being identified, but considering that the vehicle will have endured ten years of activity, the plastics will be dirty, possibly damaged and with the identification mark illegible. The solution will be to develop a hand-held tool which will 'read the code', rather like the scanner at supermarket checkouts which reads the code on each item and displays the cost. Alternatively, it may be possible to develop an apparatus which will identify the chemical composition of plastics.

At present automobiles are the principal concern of the conservation experts. The end of the Cold War has led to redundancy for military vehicles, weapons and warheads. The safe dismemberment of nuclear weapons is a highly specialized business. Perhaps we shall not 'turn swords into ploughshares', but we may expect tanks to be turned into tractors and weapons into washing machines.

FUTURE SOURCES OF ENERGY

The amount of energy required to make a metal is much more than that used in the final smelting operation. For example, bauxite ores are quarried in tropical regions, transported and refined to alumina; cryolite is manufactured and the carbon anodes produced, after which about 14 000 kWh of electric power is required to electrolyse the oxide into each tonne of aluminium.

As rich and accessible deposits of ore become depleted, more and more energy will be needed to win metals from the remaining, poorer deposits. The metal content of copper ores now mined has decreased from about 6 per cent in the nineteenth century to an average of about 0.7 per cent, and it is likely to fall to 0.2 per cent within the next few decades; the energy per tonne of copper will increase accordingly.

In addition to the energy required for winning and smelting metals, the subsequent shaping operations – casting, rolling, forging and welding – absorb precious energy. After the oil crisis in 1974, some companies did not react quickly to the immense price increases and went bankrupt. Others began to investigate possible ways of preventing the waste of energy. Often their efforts began by calculating how much energy is required, theoretically, to operate each metallurgical process. For example, from the specific heat and latent heat of aluminium it is easy to calculate how many heat units are required to bring one tonne of aluminium to the temperature at which it can be cast. One tonne would require 253 million calories, provided by 27 litres of fuel oil, 10 therms of gas or 294 kWh of electric power. One can imagine their dismay on discovering that two or three times the theoretical energy requirement was being wastefully consumed.

Many solutions were found, including waste heat recuperation to superheat incoming air at oil burners, better insulation of furnaces, automatic furnace controls to cut down the heat once the required temperature had been reached and rescheduling processes to use low cost electricity during the night. Apart from such improvements, many companies realized that energy was being wasted by sheer foolishness: one consultant found that a 20 tonne charge of molten copper was held for several hours while an overhead crane was being replaced, whereas a rescheduling of the repair would have prevented that waste of energy. Other inefficiencies, such as leaking air compressors, are surprisingly costly.

It is inevitable that our energy consumption will increase. The major supplies of coal, oil and gas are being depleted, and in different ways they all cause ecological problems. Many readers of this book will be worried about the use of nuclear energy, partly because some stations convert uranium to plutonium. There is a much-needed lull in the construction of nuclear power stations during the 1990s, giving the opportunity for realistic accounting and improving arrangements for the disposal of nuclear waste. At present about 20 per cent of the world's energy is provided by nuclear power stations, and it would have been greater had there not been the understandable fear of the word 'nuclear'. France, using pressurized water reactors and with a fine record of safety, has efficiently brought her percentage to 70 and provided a well-organized programme of publicity to the general public. Thanks to these efforts and those of Sweden, Europe's average proportion of energy supplied by nuclear stations has risen to one-third.

Nature provides some sources of energy free of charge. The Sun offers a plentiful source of energy, though it is countries with an adequate supply of sunshine and space that are most likely to benefit. The most promising method of harnessing solar energy is with photovoltaic cells, consisting of semiconductor diodes. The main problem lies in making the apparatus cheaply enough. An operating cell is expected to give about 150 W of electricity per square metre; it is evident that we must be able to produce large areas of cells very cheaply for them to become competitive. The cost has been dropping in recent years, helped by the production of amorphous silicon, which is cheaper than the crystalline variety. A great deal of research, backed by the government, is being done in the USA. Today, photoelectric cells generate electricity at a cost of about 25 cents (16 pence) per kilowatt-hour. The ultimate application, bulk electric power generation, is expected to come in the next twenty years, with a decline in cost to about 10 cents (7 pence) per kWh. The future of large scale solar energy is no longer a question of 'if', but 'when'.

Feasibility studies are being conducted to assess the possibility of extracting energy from the sea. In addition to the power from tidal energy, some small shoreline stations have been producing power from waves. In 1991 such a station was established off the Hebridean island of Islay. Waves entering a natural rock gully are channelled into a concrete chamber, forcing air through a turbine generator. Similar stations have been constructed by Norway and Japan.

Where there are high mountains and ample supplies of water, hydro-electric power is available and few processes can compete with a good dam, so countries like Canada, the USA, Norway and Venezuela are well blessed. In some places power obtained from wind is economic. This is not limited to the single windmills that produce energy for small communities. Windpower has been harnessed on a large scale in California. Hundreds of wind turbines, as high as Nelson's Column and with blades spanning 60 metres, make a beautiful sight in California's wide-open spaces, but we understand that the noise can be incessant and annoying. We wonder what Don Quixote de La Mancha would have thought of them.

These and other methods of producing energy are important, and they do not desecrate the atmosphere, but all of them have limitations since they need ample space, ample water, ample wind and ample sunshine. To produce the amount of energy equivalent to that of a typical power station, a group of wind turbines requires 500 square kilometres of open space; a solar energy station about 150 square kilometres, a tidal barrier would be 7 kilometres long and a wave-powered station about 100 kilometres long. Even if some or all of them are developed, we shall be lucky if they will cope with one-tenth of the world's energy requirements.

28

SOME COMPETITORS OF METALS

It will by now be clear that metals have remarkably useful properties. However, they do have some limitations. Metals can be made substantially stronger by alloying and heat treatment, but usually as they become stronger their ductility is reduced. They tend to soften at elevated temperatures, so that every alloy system has a temperature limit beyond which it is inadvisable to go. Some alloy systems, including those of aluminium, do not have a finite fatigue limit. A few engineering metals – magnesium, aluminium, beryllium and titanium – are light enough to be used for components of aircraft, but most other metals are heavy. Copper and gold have splendid colours and anodized aluminium can be tinted, but otherwise the possibility of producing coloured metals is limited.*

Most metals require a great amount of energy for their production and some are becoming scarce, so that alternative materials are being sought. The stage is therefore set to consider competitive materials and to describe how metallic alloys themselves can be made stronger by reinforcement with non-metallic materials.

So far the word 'composite' has been mentioned only a few times in this book. However, we have seen that the valuable properties of steel originate in the association of ductile ferrite with hard cementite or with the microscopic particles of martensite which make the steel so strong. Similarly, we have seen too how the precipitation hardening of aluminium alloys is brought about by the presence of many billions of submicroscopic particles of intermetallic compounds. These and other alloys may be regarded as materials whose composite structure has been achieved by metallurgical processing.

Timber, a composite material formed by nature, is a remarkably cheap

---

* The gold–aluminium intermetallic compound Au_2Al has a purple colour, and has been used for watch cases.

way of buying strength. For both primitive and sophisticated communities, it makes an excellent construction material. The carpenter knows how to exploit the grain of the wood to obtain maximum strength, and the DIY enthusiast uses the composite material plywood for many purposes. Timber is an ideal material for structural beams in buildings, where it is built up from separate units to achieve maximum strength. In many parts of the world, research and development is being directed to harden, strengthen and preserve timber and thus to widen its scope.

Concrete is a composite of cement and aggregate. When it is reinforced with wires or bars of steel, it becomes, so to speak, a composite composite. Reinforced concrete is the most widespread way of buying strength, and is by far the world's major tonnage material.

There are many other 'composite composites'. The modern aircraft windshield needs to be strong and light, to possess impact resistance and fatigue resistance, and to resist misting and icing. These qualities are met by a laminate containing six or more layers of glass, polymer, elastomer, woven fibre and a thin transparent metallic film.

PLASTICS

There are about 15 000 trademarked plastics available today, and it is expected that by the year 2000 the number will have doubled. In the USA, 30 million tonnes of plastics are made each year, a third of which goes into packaging. Plastics are easily fabricated and joined; they are of low density and many of them can be supplied in a wide range of attractive colours. They have good chemical resistance, but only moderate resistance to thermal degradation – a problem which causes concern because some plastics have a limited useful life.

Plastics form the largest group of synthetic materials known as polymers, in which very large molecules, some comprising thousands of atoms, have been formed by linking several groups of smaller molecules into chain or network structures. They are broadly divided into thermoplastics and thermosetting resins, the former group accounting for over 90 per cent of the plastics industry. Thermoplastics include polystyrene, high and low density polyethylene, polypropylene and polyvinyl chloride. This PVC and the polyethylenes are the most widely used plastics. There are also engineering plastics which have superior mechanical properties but cost more. Examples are polyesters and nylons, acrylics and polycarbonates. All thermoplastics melt on heating, so they can be injection moulded or extruded; they can also be blow moulded (to make plastic bottles, for example).

Injection moulded plastics are familiar in thousands of well-known shapes, ranging from the cutlery provided on aircraft to garden chairs and complete bodies of dinghies. The injection mouldings are made, usually automatically, on machines resembling those used in diecasting. The plastics granules are introduced into the cylinder of the machine where they are gradually melted by the heated barrel and the shearing action of a tightly fitting screw before being injected under pressure into a mould. The injection mouldings are accurate dimensionally and often of complex shape. After the material has been moulded, it can be recycled, provided it can be identified and separated from other plastics.

The other group, the thermosetting resins (Bakelite was an early example, and Formica a more recent one), can also be moulded, but the heat of the process changes the chemical structure of the plastics, so that they cannot be recycled. These materials are usually moulded under pressure in open moulds in a process known as compression moulding, characterized by long cycle times – often one minute, but sometimes as long as five minutes. On that account the process is not suitable for automation. Furthermore, thermosetting materials darken during the moulding process, so they cannot be offered in pastel shades.

Plastics have already made inroads into fields that were previously the province of metals. Polyethylene and PVC are used instead of copper, lead and cast iron for water conduits. Polyethylene replaces galvanized iron for buckets and other containers. PVC has replaced cast-iron guttering and domestic waterpipes; glass-filled nylon has replaced many metallic components. An automobile manufacturer states that in one of its models, out of 2730 parts 771 are made of plastic.

Although, as we remarked in the previous chapter, many more alloy compositions are still to be developed, it is not likely that many dramatic alloy discoveries will be made in the future. The molecules of which plastics are made can be joined, branched or cross-linked to form complex new polymers, so there is a chance that plastics with radically new properties will be discovered that may change the conflict between plastics and metals in favour of new materials, particularly if they have high strength at high temperatures.

The cost of plastics varies widely. Polyethylene and similar materials cost about £500 per tonne, and nylon about £3200 per tonne; the exotic polymer poly-para-phenylene terephthalamide), familiarly known as Kevlar, is manufactured in several grades and can easily cost ten times as much as nylon.

FIBRES IN COMPOSITE MATERIALS

The cables supporting a suspension bridge are assemblies of thousands of intertwined steel wires which together provide a greater strength and suppleness than would steel rods of the same diameter. When other materials are made into fibres, much thinner than the wires of suspension bridge cables, their strength per unit diameter is also greater than when they are in bulk form. Many years ago it was discovered that glass could be made stronger by producing it as a thin fibre. Most such filaments on their own are of little value as engineering materials. They are strong but brittle, so they need to be introduced into a matrix which will support them and overcome their brittleness.

Glass fibre reinforced plastics (GFRP) were first produced in the 1940s, but their great potential was not recognized. Research on these new composites continued in a rather desultory manner for another twenty years. In Britain, Rolls-Royce was one of the first organizations to appreciate the potential of such composites, but even in the 1970s they were not fully developed; now, however, they are the most widely used of all composite plastics. Most modern automobiles have GFRP body mouldings, panels and doors. They have even been used for inlet manifolds. In the home, injection moulded composites are required for kitchen equipment, furniture, and TV and computer bodies. Many pieces of sports equipment are made from these composites – tennis rackets, fishing rods, surfboards, canoeing gear and motor cycle crash helmets. The production of GFRP is a thriving international industry.

The glass fibres are made by drawing several hundred of them in parallel from a container of molten glass, through platinum dies. The thin fibres so produced are coated with a material which lubricates the contact between adjacent fibres and protects their surface. E-glass, a borosilicate composition similar to Pyrex, is the most widely used fibre material. S-glass, an aluminium–magnesium silicate, is 30 per cent stronger than E-glass but more expensive. These fibre glasses compare favourably in strength with steel piano wire: E-glass is 10 per cent stronger, and S-glass over 30 per cent stronger.

After the potential of glass fibres had been developed, a new species was discovered in the USA: boron fibres, with properties substantially better than those of any glass fibre. They are made by depositing boron on a tungsten filament, followed by a surface coating of silicon carbide or boron carbide to increase the strength between fibre and matrix and to prevent corrosion.

The matrix must hold a large number of separate fibres, aligned in the required direction and not randomly distributed. The potential for reinforcement by fibres depends on how easily they can be aligned in the polymer matrix. The aerospace industry demands strength plus lightness, and is not so constrained by price considerations as the motor industry. A variety of fibre-reinforced plastics are used in laminated panels, wing spars, fuel tanks, bulkheads and helicopter rotor blades. The Boeing Chinook helicopter contains over 20 per cent by weight of such composites.

CARBON FIBRES

Carbon exists in forms ranging from graphite to diamond, with a wide range of properties. In vitreous form it has great strength and resistance to chemical attack. Carbon fibres were first developed over a hundred years ago by Thomas Alva Edison, who used them as filaments for his first electric lamps, but in this application they were soon displaced by tungsten filaments. They are made by treating highly drawn textile fibre and stretching it while hot. They are usually woven into cloth, often of complex shape, for incorporation in the matrix of either thermoplastics or thermosetting materials.

An interesting potential application of carbon fibre is in oil platforms. Some offshore rigs stand on massive steel legs fixed to the sea bed, but such structures are only practicable for depths down to 1000 metres. For greater depths, floating platforms are attached to the sea bed by steel ropes. The weight of the steel tends to pull the platforms downwards, so the effect has to be counterbalanced by buoyancy tanks. These are practical for depths as great as 1500 metres, but beyond that the increased weight of the steel makes the tanks too large. Carbon fibres, thinner than a human hair, are embedded in epoxy resin to form ropes which are only a fifth the weight of the steel cables of equal strength. If they prove satisfactory, they can be used to attach floating platforms where the sea bed is as far down as 3000 metres. Other developments may come from this, perhaps for the cables of suspension bridges.

During the 1980s there was a continuous effort to attain maximum strength in carbon fibres, and recently what are known as 'intermediate modulus fibres' have been developed, with a higher strength and stiffness than other types of carbon fibre. All carbon fibres are comparatively expensive, but during the last fifteen years their cost has been reduced from several hundred dollars per kilogram to a few tens of dollars, and the search for still cheaper substitutes made from petroleum or coal-based pitches continues. Like other high-tech products, they have to be made under conditions of absolute cleanness.

CERAMICS

For several thousand years, refractory materials have played a vital part in metallurgical operations, for without them it would have been impossible to contain metals at the high temperatures at which they melt. Such refractories are moulded into shapes such as crucibles. A more sophisticated group, known as advanced ceramics, are characterized by an atomic structure which gives them great hardness and stability at high temperatures. Silicon nitride is a typical example in that it displays the desirable properties as well as the limitations of such materials. They are somewhat brittle, and many workers experimenting with them found that, although under laboratory conditions the ceramics behaved well, in the rougher conditions of manufacturing plants they tended to break under vibration and stress. However, such problems are being overcome, and silicon nitride has been used for turbine blades and automotive turbocharger rotors.

Ceramics such as silicon nitride and silicon carbide are capable of functioning at temperatures substantially higher than those endured by nickel alloys, so there is great scope for them as components of internal combustion engines. Insulation coatings in such materials and cermets (which combine ceramics and metals) or even wholly ceramic parts would increase the operating temperature of a diesel engine from 700 to 1100°C, which would have the effect of improving the engine's efficiency by 50 per cent. They are made by hot pressing or sintering finely powdered oxides of aluminium, zirconium, silicon or silicon nitride. Alternatively, the powders may be sprayed in a plasma directly onto a metal to provide a coating which is so hard that it can be cut only by a laser. Fibres made from ceramics also have great possibilities.

Silicon carbide has been well known as an abrasive grit and refractory material, but it was not possible to exploit its potential as an engineering ceramic until the reaction bonding process was developed in the 1960s. Silicon carbide and carbon are compacted under high pressure and then heated in a vacuum at about 1500°C in contact with molten silicon, which reacts with the carbon to form more silicon carbide. The material formed in this way is used for seals in chemical engineering, turbochargers, sea water pump bearings and gas burner nozzles.

Advanced ceramics have proved valuable to surgeons in replacing worn or damaged bones. A ceramic implant can be made porous, so that regenerated bone can grow onto the implant. The ceramic materials are light and do not corrode, which makes life easier for the patient. In 1986 such 'bioceramics' were used in over 300 000 hip operations, and it is estimated that the number will have grown to 400 000 by 1996.

METAL MATRIX COMPOSITES

Research laboratories in many countries are now producing metal matrix composites, in which a strong light alloy, usually of aluminium or magnesium, is made much stronger by fibre reinforcement. Large aluminium castings reinforced by as much as 65 per cent by volume of high strength boron fibres are being made for aerospace requirements. Titanium reinforced by silicon carbide fibres has nearly twice the strength-to-weight ratio of conventional titanium alloys.

Other composites are of aluminium alloys strengthened with fibres of silicon carbide. Recent research in the USA and the UK have led to replacing the matrix with intermetallic compounds. Thus a composite having a matrix of the titanium–aluminium intermetallic Ti_3Al and silicon carbide fibres with carbon cores is being used for turbine blades.

METALS VERSUS COMPOSITES IN BICYCLES

The bicycle has been acclaimed as the most efficient machine for converting energy into motion. Bicycles are environmentally friendly, they provide good exercise for young and old, and can weave their way through traffic jams. If the rider is expert and strong, he may win a fortune in such events as the Tour de France. Ten years from now there will still be many millions of bicycles, from Beijing to Baltimore. No doubt most of them will have frames and handlebars made of alloy steel tubes, but some may be of aluminium or perhaps magnesium alloys. Some will be very light, strong and fast because their construction, even possibly their wheel spokes, will be of titanium alloy.

The drive towards the use of composite materials such as those incorporating carbon fibre is rapidly affecting the construction of bikes for mountain use and racing. The famous 'speed merchant' Chris Boardman uses a carbon fibre machine. Many top class long distance riders use carbon fibre machines for mountain stages because they are so light, but change to titanium bikes for the other stages. There are still some problems in joining the frame tubes to the lugs with adhesives, and after a season's hard use the carbon composite frame has to be replaced. Undoubtedly this problem will be overcome during the next few years.

The bicycle wheel has been the subject of much research, and the disc wheel provides an example of the use of high strength composites. One such wheel consists of a sandwich of carbon fibre expoxy resin, with a foam polyester core. The facings are bonded to flanges, the hub is of aluminium alloy and the rim is made from a carbon fibre composite.

International races always provide interesting examples of competitive new materials. As long ago as 1974, the Spanish rider Louis Ocana rode a titanium-frame machine for some stages when he won the Tour de France. In 1980, Joop Zoetmelk won the Tour with an alloy-steel-framed bicycle, but his Raleigh machine had several small components, such as the pedal spindle, made of titanium alloy. The German rider Jens Gluchlich had many successes with a machine that had its frame, handlebars and disc wheels made from carbon fibre composite. We gather that Miguel Indurain, winner of the 1993 Tour de France, prefers alloy steel frames.

Many years ago, one of us suggested that in the future there might be a concrete bicycle; this caused much hilarity. Since then there have been impressive developments in fibre-reinforced concrete, and we understand that an experimental concrete bicycle has been made. The first plastics-and-composite bicycle was made in Sweden. Its wheels were of a glass fibre reinforced composite, and the frame was made from an injection moulded glass fibre reinforced polyester. In Japan, a super-lightweight rideable bicycle has been made of paper.

29

METALS AND MATERIALS AS A CAREER

A generation ago, metallurgists were more or less confined to their laboratories, so opportunities for advancement were not plentiful. The general public had little understanding of the functions of metallurgists; indeed, many people thought they had something to do with weather forecasts. When a class of children was asked to define a metallurgist, one of them hazarded the guess that it was 'a small animal, nice and furry'. The importance of the subject is now more widely recognized, and metallurgy as a career offers the prospects of managerial positions, instead of the role of advisory expert. The study of metals has been broadened. University departments which used to specialize in metals now embrace other materials, including the ever-increasing range of composites discussed in the previous chapter.

PREPARATION FOR FURTHER EDUCATION

Nowadays information is available to students seeking careers in metallurgy and materials engineering. For example, the UK's Institute of Materials has produced publications for schools, covering subjects that would appeal to young people, such as the design of bicycle frames – comparing one made of metal with another in a polymer-reinforced composite; also tennis rackets and the materials for a compact disc player were discussed.

When the choice of university is being considered, the would-be student will receive illustrated brochures, not only describing the materials engineering course, but providing helpful information about accommodation, and facilities for sport, music and social events. Most universities invite young people to look around the department, meet the staff and get to know others who are considering coming to the university.

University courses are designed to make students aware of the opportunities and responsibilities of management. It is not uncommon for students to work for several months with manufacturing companies. Their duties are discussed with the university staff, the employers and the student. On return he or she will have a better idea of 'what it's all in aid of', and will have observed and analysed some of the problems they will come up against in their future careers. The study of management, including cost accountancy, is entering more and more into university and college courses. Students are now required to design plant layouts for manufacturing processes, to estimate the cost of building, power and materials and to link their studies of the project with visits to plants which are involved with similar problems.

Managers must be concerned with conservation of energy and materials, and with environmental issues and safety in the workplace. One regrettable cause of accidents in the metallurgical industries has been the movement of hot or molten metal. An ever-present worry for those who are training young workers is the danger of pouring molten metal into a mould which is moist. Only rigorous instructions and insistence that safety glasses and protective clothing are worn will eliminate such accidents. Safety starts with the engineers who design guards and other equipment, and finishes with the safety committee which makes sure that the equipment is used intelligently; however, it is also essential that managers give full backing to these endeavours. We must also stress that, having entered industry, the manager-to-be should take every opportunity for further training in financial management and industrial relations. Now mathematics, electronics and other disciplines are associated with computerized and automated manufacturing processes and a vast range of new materials have become important.

Great opportunities exist for women. No doubt a generation ago the heat of the processes, the danger of splashing metal and sometimes the 'macho' attitude of male workers made women feel out of place in the 'man's world' of metallurgy. When the authors were students in the 1930s, just one shy young lady became a metallurgy student, and she remained for only two years. Women now account for a substantial intake of the departments of metals and materials.*

While discussing opportunities for management, we must emphasize the need for fluency in at least one foreign language, now that international trade is an essential feature of industry. Many US and UK managers find that executives in other countries speak perfect English and negotiate in that language. But during the course of contact it is

---

* We are confident that we will be forgiven for not retitling this book *Metals in the Service of Persons*.

usually essential to have discussions with works personnel, who appreciate a conversation in their own language, provided it is fluent and colloquial. Many English-speaking companies are now engaging graduates from other countries because they are technically qualified and fluent in English plus one or two other languages.

There are many opportunities for interesting and rewarding careers, and in this chapter we discuss some of them, ranging from mining metal ores in remote parts of the world to specialized work in a research laboratory. In contrast to the conditions of several decades ago, co-ordination is necessary. The research scientist will inevitably be required to spend time in the production plants of the same group, to discuss problems and possibilities of new discoveries.

MINING AND PRODUCTION

During the last twenty years the costs of many materials have increased substantially, while rich ores of some metals have become scarce. These factors led to the exploitation of new deposits, a reappraisal of the possibilities of weak ores, and the development of new extraction processes suitable for those ores. Fortunately the increased value of some by-products has a useful effect on the profitability of smelting processes. For example, in 1992 Kennecott's Bingham Canyon mine (described on page 14) produced 300 000 tonnes of copper, but also 10 000 tonnes of molybdenum, 130 000 kg of silver and 15 000 kg of gold. In addition to the production at Bingham Canyon, Kennecott has other mines in Utah, Nevada, South Carolina and Alaska. The total output of the by-product gold makes them one of the world's largest producers of it.

The efforts to produce metal profitably have been linked with new methods of discovering ore deposits, more efficient methods of excavation, transport and concentration of ores, and improved methods of smelting those ores with less energy than before. Such endeavours, coupled with the vast increase in the recycling of scrap metals, will become one of the biggest challenges of the future.

METALS AND MATERIALS IN THE CONSUMER INDUSTRIES

Several metals once classified as exotic rarities are becoming widely used: samarium, praseodymium and neodymium in magnetic alloys, and hafnium and gadolinium in nuclear engineering. The separation, alloying, properties and fabrication of such metals need to be studied extensively.

Materials, including composites, with the greatest possible strength combined with resistance to heat and minimum weight are required for aircraft and space vehicles. At the other end of the scale, as we saw in an earlier chapter, the manufacture of such simple articles as a soft-drinks can or a needle involves continual research and development.

THE MATERIALS ENGINEER AS SHERLOCK HOLMES

Investigation of aircraft and motor car accidents or bridge collapses involve retesting and microstructural examination of parts and much work of a forensic nature. An important spin-off is that examinations of failed structures increase our total experience and knowledge, leading to improved design, manufacture and construction and hence prevention of failure. If the effect of the harmonic oscillations that destroyed the Tacoma Narrows Bridge or the potential weakness of box girder design had been foreseen, much suffering and waste of money would have been avoided. Metal fatigue is an ever-present hazard, especially when a structure operates in corrosive conditions. The Norwegian oil platform *Alexander Kiellan* suffered a catastrophe which led to the loss of over a hundred lives, because what was thought to be a minor modification was undertaken without considering its possible effect. The platform was supported on vertical legs, held by steel tube braces about a metre in diameter. It was decided to drill a hole in one of the braces so that some communication equipment could be fitted; after that the brace was welded. It subsequently developed a crack which grew to over a metre long, causing the collapse of the brace and then the whole structure.

Aircraft failure investigation represents an important test for the metals detective. Sometimes a small error can lead to a major disaster. In 1988 a Boeing 737 caught fire at Manchester Airport, with the loss of many lives. The accident was caused by fatigue in a pressure chamber causing fuel to ignite. During the investigation it was discovered that a welding repair had been performed and, though the work conformed to legal requirements, it had been optimistically assumed that the repaired component would have the same service life as a new one.

RESEARCH IN METALS AND MATERIALS

Although in the past great scientific discoveries were made using primitive equipment, present-day scientists are provided with measuring instruments of almost unbelievable accuracy. It is possible to measure time intervals of less than a picosecond – the time it takes for a ray of light to travel one-third of a millimetre. We can now observe the processes of chemical change while they are taking place. When the authors were doing research in the 1930s, their microscopes gave a magnification of about a thousand, so that, for example, the mechanism of age hardening could not be explored. Now the electron microscope and other instruments mentioned in Chapter 6 have made it possible to isolate and examine the behaviour of minute groups of atoms and molecules.

A great deal of pure research is done with the aim of extending our knowledge, often without any forecast of industrial developments it might bring. Other research is carried out to further the use of new discoveries or to improve existing processes. Many such efforts have been mentioned in previous chapters, ranging from the study of the crystal structure of metals to the wonderful progress in the steelmaking industry, from the Bessemer converter to the modern oxygen furnace.

J. Robert Oppenheimer once said, 'Every new finding is a part of the instrument kit of the scientist for further investigation or for penetrating new fields.' The research scientist has therefore to be alert to the possibilities of new discoveries. A French metallurgist is probably kicking himself in Purgatory. His name was Berthier, and in 1820 he observed that when chromium was alloyed with steel the metal became resistant to corrosion. He did not appreciate the significance of his discovery, and it was not until 93 years later that Harry Brearley discovered stainless steel.

It is important for metallurgists and materials scientists to keep abreast of developments in other disciplines. Fortunately, information technology now makes it possible to examine the results of a huge amount of research. Fifty years ago, scientists had to wade through masses of books and papers in the hope of finding something relevant to their own researches. Now each technical library has computers capable of displaying information about all previous researches on the subject being investigated.

More and more, the various disciplines are becoming interlinked. The recent discoveries in superconductivity illustrate how physics, chemistry, electrical engineering and metallurgy can combine with dramatic effect. In March 1987, three thousand physicists crammed into a 1200-seat auditorium in New York; the meeting began at 7 p.m. and finished at dawn. Many talks were given, centred on the work of two Swiss researchers who had discovered new superconductive materials and who later won a Nobel prize. Superconductivity had been discovered in 1911, when Heike Kamerlingh Onnes succeeded in liquefying helium, and went on to test the electrical resistance of mercury at lower and lower temperatures. He was amazed that at a few degrees above absolute zero the metal suddenly had no resistance. Then he found that lead became superconducting at 7 degrees above absolute zero. In subsequent years, other superconducting materials were found, including niobium–tin and niobium–titanium, mentioned on page 212, but these required very low temperatures – 18 K and 9 K respectively – obtained by the refrigerating effect of costly liquid helium. The Swiss discovery had been published in 1986 but was little noticed until the results were confirmed by American and Japanese scientists. By March 1987, hundreds of researchers were working on superconductors and many of them clamoured to establish their claims at the famous meeting.

The newly discovered materials belong to a group known as perovskites; typical ones are ceramics containing copper, each copper atom being surrounded by six oxygen atoms, with either barium plus lanthanum, barium plus yttrium or bismuth plus strontium and calcium. One of the 'high temperature' superconductors, based on bismuth, is capable of operating at 110 K and another, based on thallium, is superconducting at 125 K (minus 148°C). Such perovskites are refrigerated with liquid nitrogen, which is suitable for temperatures over 80 K. It is easy to use and is inexpensive, costing about the same as petrol. Liquid helium is much more costly and, in spite of its low temperature, is far inferior to liquid nitrogen in its ability to transfer heat.

One of the present problems is that the superconductors are brittle. However, there have been many examples of the successful 'taming' of intractable materials. For example, tungsten seemed an impossible metal for the filaments of electric lamps because of its brittleness, but finally it was formed into the strong flexible filaments which feature in the electric lamps of today.

The brittleness of perovskites is being overcome by supporting them in a matrix, reminiscent of the way in which glass fibres, discussed in the previous chapter, are supported. In a typical process the perovskite powder is packed in a silver tube, drawn to a round wire, rolled into a flat strip and sintered. Such a material can be produced economically, and so far lengths of over 100 metres have been manufactured. A great deal of development has been done in Japan, America and Germany; it is pleasing to report that Britain has made important progress in bringing superconducting materials into commercial use.

Apart from demonstration kits such as 'floating magnets', an early application of the perovskites was a device known as the Josephson Junction, a superconductor–insulator–superconductor sandwich which can 'switch' much faster than semiconductor switches, thus helping to make faster computers. The next five years should see valuable developments in high current conductors, super-high-field magnets and transformers. Superconductors repel magnetic fields, so they will stay suspended above a magnet. Already, levitation trains are under development. In 1990 the Japanese government authorized the construction of a 42 km prototype levitation line which will test trains running at over 500 kph.

RESEARCH AND DEVELOPMENT IN THE FUTURE

Designing for a combination of the well-established engineering properties – strength, hardness, fatigue and wear resistance, forming properties and corrosion resistance – has begun to be understood, exacting though it is. But to add other demands like specific electronic, magnetic and optical properties makes the task of engineers and merallurgists highly complex.

Research on new processes is always stimulating, though often harassing, when a new discovery that promised much in the laboratory takes a long time to reach full-scale production. The state of mind needed in research is what we might describe as 'the man from Mars' approach. Such a superman might visit a terrestrial manufacturing company and, as he had no preconceived ideas, might notice the illogicalities or follies of the process. In research on new developments there are at least three stages, each of them needing this 'man from Mars' attitude. First, the process is worked out in the laboratory. Next, a medium-sized pilot plant is built, operating in the works, but still under careful control and measurement. Finally the large capital expenditure for a full-scale plant is authorized. As production increases new problems arise, and often efforts must be made to gain the enthusiasm of the works departments, whose members sometimes have an unsympathetic attitude to the so-called 'experts' in the research laboratory. There are great opportunities for young and clever scientists, engineers and metallurgists who realize that, apart from devising a new process, the perfecting of it requires unmitigated attention, coupled with a measure of diplomacy. To quote Professor Oppenheimer again, 'Discovery follows discovery, each both raising and answering questions, each ending a long search and each providing new instruments for a new search.'

GLOSSARY

(Technical terms which are indexed and amply defined in the text are not included in the glossary.)

AGEING As applied to castings in steel and cast iron, the word indicates a period of time provided to relieve casting stresses. Ageing is also used in reference to wrought aluminium alloys of special composition; after heat treatment they are quenched in water and then kept at room temperature for some days. During this period, their maximum strength and mechanical properties develop fully.

ALPHA PARTICLE The nucleus of the helium atom, containing two neutrons and two protons, emitted from the nuclei of certain radioactive elements.

ANGSTROM UNIT A unit of measurement used by metallurgists and crystallographers, giving the distance between atoms in a space lattice. An angstrom unit is one ten-millionth of a millimetre. See page xvii.

ANNEALING The process of heating a metal or alloy to some predetermined temperature below its melting point, maintaining that temperature for a time, and then cooling slowly. Annealing generally confers softness.

BILLET A bar of metal, usually steel, made as an intermediate shape when converting ingot into strip or rod.

BINARY ALLOY An alloy composed of only two major constituents, e.g. copper and zinc, or lead and antimony. An alloy with three constituents is known as a ternary alloy.

CATALYST A substance which, when present in small amounts during a chemical reaction, promotes the reaction but itself remains unchanged.

CEMENTITE The name given to identify one constituent in iron–carbon alloys. Cementite is iron carbide, Fe_3C, but may contain other elements such as manganese and chromium, carbides of which are dissolved in the iron carbide. Its name was coined in the 1880s by Professor Sorby, who said it was 'iron with cement carbon'.

CERMETS Materials produced by bonding a metal oxide, carbide nitride or boride with a ceramic. The bonding is effected at high temperature under controlled conditions by methods similar to those used in powder metallurgy.

COMPATIBILITY Two materials are said to be compatible when they can exist in contact with each other without interaction.

CORES Specially fashioned pieces of sand or metal used to form the hollow parts of a casting. To make a cylindrical hole in a casting, a cylindrical solid core of similar shape is used.

CRYOGENIC The condition of materials and the behaviour of phenomena at very low temperatures.

DEEP DRAWING A shaping process in which the whole or part of a disc of metal is forced through the aperture of a die so as to make a cup shape. By repeating this process with plungers and apertures of progressively decreasing diameters, the metal is drawn into an elongated cup or closed tube.

DISTILLATION The conversion of the whole or part of a liquid substance into gaseous form to be followed by subsequent condensation to liquid.

DOPING The addition of minute amounts of another element to a semiconductor material, to bring about an immense change in its electrical conductivity. A typical dopant is boron added to silicon, in the proportion of between one and a hundred parts to a million.

DUCTILITY The property of a metal which enables it to be given a considerable amount of mechanical deformation (especially stretching) without cracking.

ELECTRODE A conductor which conveys electric current directly into the body of an electric furnace, plating vat or other electrical apparatus.

ELECTRON Elementary, negatively charged particle having a mass about 1/1840 that of a hydrogen atom.

ELEMENTS All compounds can be resolved into elements: for example water into hydrogen and oxygen; common salt into sodium and chlorine. The elements, however, cannot be resolved by chemical means into any simpler substances. Including the modern 'synthetic elements' which have been made in the atomic physicist's laboratory, there are over a hundred elements, of which more than three-quarters are metals.

EQUILIBRIUM The state of balance which exists or which tends to be attained after a chemical or physical change has taken place. Equilibrium may not be reached for long periods after the change has been initiated.

FLASH When a metal is forged or cast in a die, some metal penetrates the space where two die surfaces touch and thus a web of metal, known as 'flash', remains attached to the forging or casting and has to be removed by filing or clipping.

FLUX A chemical used to combine with a substance having a high melting point, generally an oxide, forming a new compound which can readily be melted.

GAMMA RADIATION Rays of very short wavelength, less than that of X-rays, produced during the disintegration of radioactive materials.

HOT SHORTNESS An undesirable property of certain metals and alloys whereby they are brittle in some high temperature range.

HYPERBARIC WELDING A method of producing welds underwater in a dry habitat at greater than atmospheric pressure. The further below the surface the position where the weld has to be made, the higher the pressure in the working chamber, since the pressures must be sufficient to displace water within the chamber.

INDUCTION FURNACE The metal is held in a refractory container surrounded by a coil through which alternating current is passed. This induces currents in the metal, causing it to be heated and, if required, to melt by internal resistance. The frequencies of the applied currents vary according to the amount of metal. Small quantities in laboratories are melted in high frequency furnaces ranging from 10 000 to a million cycles per second. Medium weights of metal, for example half a tonne of aluminium alloy melted in a foundry, require a frequency of 1000 cycles. Very large tonnages are melted in low frequency furnaces operating at 50 cycles per second – the same as that used in the UK for domestic supply.

ION An electrically charged atom or group of atoms.

ISOTOPES Atoms of the same element, having the same number of electrons and protons but different numbers of neutrons. The isotopes of an element are identical in their chemical and physical properties except those determined by the mass of their atoms.

LATERITIC ORES The word is derived from the Latin for 'brick'. Red-coloured clay-like laterite materials are used for road building in the tropics. Deposits of lateritic ores, containing small amounts of nickel, are found in equatorial regions. They contain much moisture and compounds of iron which have to be removed, with difficulty, before the ore is in a suitable condition for smelting.

MACH NUMBER The relation between the speed of a moving body to the local speed of sound, which can vary with temperature, altitude and, therefore, pressure. The term is generally used in connection with the speeds of supersonic aircraft. Mach 1 is the speed of sound at sea level; the speed of sound is reduced with increasing height. The word is derived from the name of Ernst Mach, a nineteenth-century mathematician and philosopher.

MAGNETIC PERMEABILITY The ratio of the strength of magnetism in a material to the strength of the external magnetic field which induces it.

MALLEABILITY A property of metals enabling them to be hammered and beaten into forms such as that of thin sheets, without cracking. Gold is the most malleable of all metals.

MICROMETRE A millionth part of a metre: a thousandth part of a millimetre, 10 000 angstrom units. Formerly known as a micron.

MODULUS OF ELASTICITY The ratio of stress to strain in a material. The strain is usually a measure of change of length. The stress is a measure of the force applied to cause the strain. (*See* Young's modulus.)

NEUTRON An electrically uncharged particle possessing a slightly greater mass than the proton. Neutrons are constituents of all atomic nuclei except that of the normal hydrogen atom.

PROTON A positively charged particle in the nucleus of an atom having a mass about 1840 times that of the electron and an electric charge equal in magnitude to the negative charge of the electron.

REFRACTORIES Firebricks or other heat-resisting materials used for lining furnaces and retaining the heat without allowing the outer shell of the furnace to be damaged. Refractories are classified as 'acid', 'basic' or 'neutral' according to their composition and their action on the hot substances with which they come into contact in the furnace. Examples of the three types are silica, dolomite and carborundum, respectively.

RETORT A specially shaped vessel for containing substances which are to be heated to form a vapour, which is then collected or condensed.

ROASTING The process of heating an ore at medium temperature in contact with air.

SEMICONDUCTORS Materials which at room temperature have much lower electrical conductivities than metals, but whose conductivities increase substantially with increase of temperature. This is in contrast to the electrical behaviour of metals, whose conductivities decrease slightly with increase of temperature. Silicon, germanium and selenium are semiconductors. They derive their importance from the fact that their conductivity is extremely sensitive to the presence of impurities. Almost all electronic devices in use at present include semiconductors.

SLAGS Glass-like compounds of comparatively low melting point, formed during smelting when earthy matter contained in an ore is acted on by a flux. If the earthy matter were not deliberately converted into slag, it would clog the furnace with unmelted lumps. The fusibility and comparatively low density of the slag provide a means by which it may be separated from the liquid metal.

SOLUTION The intermingling of one substance with another in so intimate a manner that they are dispersed uniformly among each other, and cannot be separated by mechanical means. Although solid substances dissolved in water are the most familiar solutions, the word has a wider meaning: for example, gases dissolved in liquids, liquids dissolved in liquids and, frequently in metallurgy, solids dissolved in solids are all forms of solution.

SPECIFIC GRAVITY The ratio of the density of a substance to the density of water.

SPECTRUM The result of resolving radiation into its constituent wavelengths. The spectrum of white light, consisting of coloured bands, is the most familiar, but all types of radiation comprising a range of wavelengths can be resolved into spectra.

TEMPERING A warming process intended to alter the hardness of a metal which has already been subjected to heat treatment. The tempering temperature is lower than that at which the first heat treatment was carried out.

THERMIT A mixture of powdered iron oxide and aluminium which, when ignited, sets up a vigorous chemical action whereby the aluminium unites with oxygen from the iron oxide, leaving metallic iron. A high temperature develops, so great that the iron is melted.

THERMOCOUPLE When two wires of different metals are joined at each end and one junction is heated, a small electric voltage is produced which depends on the temperature and which can be measured by a delicate instrument. Certain combinations of metals are specially suitable, by virtue of the amount of current produced and their resistance to heat. Thermocouples are used for some types of pyrometer.

TROY A system of weights used for gold and silver. A kilogram contains 32.15 troy ounces.

ULTRASONIC Pressure waves of the same nature as sound waves but whose frequencies are above the audible limit.

X-RAYS Radiation of a character similar to that of light, but of a much shorter wavelength and possessing the property that it can pass through opaque bodies.

YOUNG'S MODULUS The ratio of applied stress to the strain which occurs when a metal is subjected to tension or compression (also known as the modulus of elasticity). It was named after Thomas Young (1773–1829), a genius in medicine, physics, optics and Egyptology.

ZONE REFINING A method of purifying crystalline materials. A bar of a solid substance is progressively moved through a furnace in such a way that there is a small molten zone; during subsequent solidification some of the impurities diffuse towards the portion last to solidify. By repeating this operation progressively the impurities become segregated at one end of the bar, which is then removed. Metals can be zone refined to contain less than one part in a million of impurity; germanium has been produced with only one part of impurity in a hundred thousand million. The technique can be used for non-metallic materials, and is an essential part of the manufacture of the ultra-pure silicon used in electronic devices.

INDEX

Abbeydale industrial hamlet 78
acid steel 117
acoustic emission 103
actinides xii
adhesives 244
advanced gas-cooled reactor 257
age hardening 51, 157
aircraft 1, 106, 160, 195, 199, 208, 282, 292
Alcan 33, 165
alchemy 2
Alcoa process 35
Alexander Kiellan (oil platform) 292
alloys 37, 93, 268
   atomic structure 58
   corrosion 228
   manufacture 38
   mechanical properties 60
   melting points 38, 39
   metallography 49
   superplastic 42
   see also under individual metals
alloy steels *see* steels
alpha particles 252
alumina 31
aluminium 30, 150
   alloys 37, 50, 195, 270
   cans 163
   casting 154
   conductors 152
   joining 236
   melting 74
   mirrors 154
   ores 10, 31
   oxide 151, 228
   production 30
   properties 151
   radiators 165
   space frame 165
   uses 153, 160, 162, 164
   welding 240
   wrought 155
aluminium bronze 41, 69, 173
amalgams 210, 221
amorphous metals 56, 273
Anaconda mine 168
annealing 133
anodes 32
anodizing 151
antimony 201
Appert, Nicholas 175
arc welding 239
armour 235
armour plate 190
arsenic 201
Aston, Frederick 251
atomic bomb 254, 256
atomic weight 250
atoms 56, 250
Audi cars 165, 267
austenite 132
Australia 11, 31, 183
Austria 118
automation 74, 79, 125, 135, 163, 176, 242
automobiles 37, 89, 127, 147, 164, 181, 196, 214, 267, 275

Backbarrow charcoal furnace 18
bacteria 16

barium   201
basic oxygen furnace   119
basic steel   117
batteries   183, 203, 209, 210
bauxite   *see* aluminium ores
Bayer, Karl Joseph   32
Bayeux Tapestry   235
bearing metals   173, 177, 207
beer barrels   143
Bell, Alexander   168
bells   93, 172
beryllium   202
Bessemer, Henry   3, 115
Bessemer converter   115
bicycles   84, 196, 236, 287
Bingham Canyon copper mine   13, 291
biotechnology   16
bismuth   41, 202
Black Country Museum   237
blast furnaces   18, 22, 25
body centred cubic lattice   58, 130
Boeing aircraft   161, 199
Bohr, Niels   266
Bonawe furnace   18
boron   139, 202
boronizing   136
Boulton, Matthew   92, 93
Bragg, Lawrence   62
brass   38, 58, 171
Brazil   10, 18, 212
brazing   165, 236
Brearley, Harry   141, 293
bridges   1, 89, 105, 113, 116
Brinell hardness tester   101
Britannia silver   220
British Steel   19, 82, 89, 127
bronze   2, 71, 94, 167, 172
Brunel, Isambard Kingdom   114
bubble-raft experiment   62
Buckyballs   271
building, use of metals in   143, 162, 180, 185
buttons   90

cadmium   170, 203
caesium   203
calcium   204
californium   xii, 263

Canada   29, 187, 193, 215, 224
cans
  aluminium   163
  nuclear   257, 264
  'tin'   176
capacitors   221
carbon   252, 271
  in steel   128
carbon fibres   285
carbonyl process   188
cartridges   90, 171, 231
case hardening   135
cast iron   17, 113, 146, 191
  grey   148
  microstructure   147
  uses   147, 165
  white   148
  see also malleable cast iron, spheroidal graphite iron
casting   65
catalysts   202, 214, 216
catalytic converters   214, 217
cathodic protection   229
caustic cracking   230
cavitation erosion   232
Celsius, Anders   xvii
cementite (Fe_3C)   45, 130, 133, 147
centrifugal casting   77
centrifuges   255
ceramics   286
cerium   216
chains   237
chain reaction   252, 254
chain-saws   196
Champion, William   178
Channel Tunnel   1, 170, 223
charcoal iron furnaces   18
Charpy impact tester   101
Chaucer, Geoffrey   3
China   167, 266, 268
chlorine   199
chromium   204
  in cast iron   148
  plating   189
  in steel   142
Churchill, Winston   231
clad metals   160, 228
Clarke, F. W.   8

coal, as substitute for coke   28
coated steel   126
cobalt   204
coinage   1, 92, 220
coke   21
cold curing resins   68
cold fusion   213
cold rolling   83
Colossus of Rhodes   167
columbium   *see* niobium
composites   281
computer control   27, 120, 122, 135
Concast process   125
concentration of ores   14, 32
Concorde   4, 157, 199
concrete   282, 288
'condenseritis'   199, 231
condenser tubes   193
conductivity, electrical and thermal
    152, 170, 226
conservation   274
consumable-arc furnaces   145
continuous casting   77, 82, 93, 123
control rods (nuclear)   257, 265
Cooke, William   170
copper   16, 167
    alloys   1, 93, 169, 233
    conductors   170
    ores   13, 14, 168
    properties   169
    uses   170
    see also aluminium bronze, brass,
        bronze
cored structure   53
cores, in casting   68, 73
corrosion   160, 196, 226
corrosion fatigue   106
corrosion tests   233
Cosworth casting process   68
crack detection   102
creep   106
critical mass   256
Cronstedt, Axel   187
crucibles   65
cryogenic steels   190
cryolite   31
crystal structure   56
crystallites   109

cupolas   146
cupro-nickel   94
Curie, Marie   215
curtain walling   162
cutting tools   200, 221, 248
cyanide
    in gold mining   206
    in heat treatment   137
cylinder blocks   74, 164
cymbals   172

Daly, Marcus   168
Darby, Abraham   18
darts   223
Davy, Humphry   4
deep drawing   90
dendrites   53, 62
Densimet   223
dental alloys   210
desalination plant   193
deuterium   213, 251
Deville, E. H. St Claire   30
Diane de Gabies   30
diecasting   73, 164, 181, 196
differential aeration   228
diffusion bonding   241
direct reduction of iron ore   28
dislocations   60, 109
dolomite   11, 21, 25, 197
doping   201, 205, 219
Dow Chemical Company   197
drop forging   79
dual-phase steels   144
Duralumin   157
Durand, Peter   175
Dyfi 'Ffwrnais'   18
dysprosium   216

Edison, Thomas A.   168
Egypt   xv, 2, 176, 206
Ehrhardt process   85
Eiffel, Alexandre Gustave   114
Eiffel Tower   228, 235
Einstein, Albert   254
elastic limit   99, 108
electrical conductors   152, 170
electric furnaces   65, 120
electric lamp filaments   223, 274

electric vehicles   184
electroforming   189
electrolytic smelting   33
electrons   250
electron beam heat treatment   138
electron beam welding   240
electron microscopes   53, 158
electroplating   181, 189, 204
elongation   100
endurance limit   105
energy sources   278
environment   35, 184, 214
equilibrium   51
erbium   216
Eriez separation process   276
Eros, statue of   151
erosion   232
etching   45, 46
europium   216
eutectics   39, 52
eutectoids   130
extraterrestrial sources of ore   268
extrusion   79

face centred cubic lattice   57, 130
Fairbairn, William   104
Faraday, Michael   168
fast breeder reactor   220, 260
fatigue   103
Fermi, Enrico   253
ferrite   132
ferrites   201
ferritic nitrocarburizing   138
ferro-alloys   13, 38, 209
Ferrostan process   176
Fiat   241
filaments   223, 274
fish farms   193
fission   253
floating magnets   295
Florence   167
flotation process   15
fluidized beds   137
foil   153
Ford Motor Company   69, 165
forging   78
Forth Bridge   235
foundries   66

fracture mechanics   110
France   31, 260, 279
freezing range   39
French Revolution   93
friction welding   241
Frisch, Otto   253
fullerenes   271
fusible alloys   41, 207

gadolinium   216, 265
gallium   205, 293
galvanic corrosion   152, 227
galvanic series   227
galvanizing   180, 229
gangue   10
gas centrifuge   255
gas-cooled reactors   257
gas cylinders   85
gaseous diffusion   255
gas turbine engines   192
General Motors   164, 184
germanium   205
Germany   145, 157, 177, 194, 275
Ghiberti, Lorenzo   167
Gilbert, Alfred   151
glass   178, 211, 214
glass fibres   284
gold   7, 206
   alloys   206
   carat values   207
   leaf   206
   mines   13
   properties   206
   rolled   207
grain refining   145, 195
grain structures   45, 52, 56
Gramme, Zenobe   30, 168
gravity diecasting   73, 74
Great Britain (ship)   114
gun metal   173

Hadfield, Robert   140
hafnium   207
Hahn, Otto   253
Hall, Charles   30
hallmarking   206, 220
halogen process (tinplate)   177
Hammersley Iron Province   11

Hardenable silver   220
hardener alloys   38
hardness testing machines   101
Healey, Michael   168
heat regeneration   279
heat resistant alloys   191
heat treatment   133
heavy water   251
hematite   19
Héroult, Paul   30
hexagonal lattice   58
high speed steels   143, 211, 222
high temperature treatment   136
holmium   216
honeycomb technology   161
hot blast furnaces   23
hot box moulding   68
HSLA steels   144, 224
hydroelectric power   34, 278
hydrogen   195, 213, 250
HYL process   28
hyperbaric welding   242
hypereutectic alloy   164

impact extrusion   90
imperfections in crystals   60
Imperial Smelting process   179
incendiary bombs   194
indium   207
Industrial Revolution   3, 168, 235
injection moulding   283
integrated circuits   219
intermetallic compounds   50, 281, 287
International Standards Organisation   37
Invar   192
investment casting   75
iridium   208
iron   3, 16, 113, 227, 266
   lattice structure   58, 130
   metallography   45, 131
   ores   11, 19
   production   17
   see also cast iron, steel
Ironbridge   113
isostatic pressing   247
isotopes   209, 250, 273

Jamaica   31
Japan   27, 138, 145, 221
jewellery   75, 206, 214
joining processes   235
Josephson junction   295

Kammerlingh Onnes, Heike   294

ladle metallurgy   122
lambda-sensors   225
Langer, Carl   188
lanthanides   xii, 216
lanthanum   216
lasers   139, 170, 225, 255, 272
laser welding   273
lateritic ores   188
lattice structures of metals   57
Laue, Max von   56
LD process   118
lead   182
leaded brass   171
leaded bronze   173
light-bulb filaments   223, 274
limestone flux   21
lithium   209, 267, 270
locomotives   266
lost wax casting   71, 75
low expansion alloys   192
low melting point alloys   41
low pressure diecasting   74
low temperature treatments   137
lubrication   212
lutetium   216

machining of metals   90
magnesite   11, 197
magnesium   194
   alloys   50, 195, 257
   ores   11, 197
   production   194, 197
   properties   194
   uses   196
magnetic alloys   203, 270
magnetite   19
Magnox   257, 264
Malaysia   14
malleable cast iron   149
management   127, 290

manganese   209, 266
manganese bronze   210
manganese steel   140
Mannesmann process   85
maraging steels   190
Marcona process   19
marine fouling   233
Marinelli bell foundry   172
martensite   134
Martin, Pierre and Emil   116
Mary Rose (ship)   200
mass spectrograph   251
Maugham, Somerset   266
maundy money   220
Mazak   182
Meitner, Lise   253
melting points of metals   17, 38, 39, 40,
    41, 175, 183, 188, 202, 203, 205, 208,
    210, 211, 212, 214, 217, 220, 221,
    222, 223
memory chips   219
mercury   210
metal inert gas welding   240
metal matrix composites   287
metallization   181
metallography   4, 44
metalloids   5, 201
meteorites   7
metrication   xv
Mexico   28
microalloyed steels   145
microbiological corrosion   233
microscopes   44
Midrex process   28
mineral nodules   267
minimill steelworks   126
mining   13, 291
misch metal   216
moderators   256
molybdenum   211
Mond, Ludwig   188
Monel   193
motor cars   *see* automobiles
moulds   66, 69
Mushet, Robert   222

Napoleon Bonaparte   175
needles   64

Neilson, James   23
Nelson, Lord Horatio   168
neodymium   216
neptunium   215
neutrons   252, 257, 263
New Caledonia   187
New Zealand   19
newton (force unit)   97
nickel   187
    alloys   191, 193
    ores   187
    production   188
    properties   188
    steels   190
    uses   188
nickel cast iron   191
nickel silver   193
nickel steels   190
Nimonic alloys   191
niobium   145, 212
nitriding   137
Nobel, Alfred   161
non-destructive testing   102
non-metals   5
normalizing   135
nuclear energy   256, 279
nuclear fission   256
nuclear reactors   224, 256

oil platforms   127, 143, 163, 226, 242,
    285
open hearth process   116
Oppenheimer, Robert   293, 295
ores   8, 11
ore carriers   19
osmium   212
oxy-fuel welding   239
oxygen
    in cast iron founding   146
    in steelmaking   118

Pacinotti, Antonio   168
Pacz, Aladar   155
paint   186, 200, 229
palladium   213
Parthenon   200
pattern welding   237
pearlite   132

Peary, Robert 7
perovskites 294
pewter 177
phases 60
philsophers' stone 2
phosphor bronze 172
phosphorus in steel 115, 119
photography 220
pig iron 114
pilgering process 87
Pilkington process 178
pipelines, gas and oil 224
Pisa, Leaning Tower of 186
Pisa bronzes 167
Pisano, Andrea 167
pitchblende 224
plasma arc welding 241
plastics 276, 282
platinum 214
plutonium 214
point defects 60
polonium xii, 264
pore-free diecasting 76
post-irradiation examination 265
potassium 215
powder metallurgy 246
praseodymium 216
precipitation treatment 159
pressure diecasting 73
pressurized water reactors 260, 265
printed circuits 177
promethium 216
proof stress 101
propellers 69
protons 250
Pu-Abi, Queen 206
pyrometers 120, 214
pyrophoric alloys 216

quality control 110, 127, 182
Queen Elizabeth 2 (ship) 162, 240
Queen Mary (ship) 235, 240

radiators 165
radioactivity 215
radium 214
railways 82, 211, 266
rapid solidification 273

rare earth metals 215
recrystallization 47, 62
recycling of metals 274
refractories 27, 120, 286
refrigeration of mines 13
reinforced concrete 282
reverberatory furnaces 65
rhenium 217
rhodium 214, 217
Riley, James 190
risers (in casting) 66
rivets 235
Roberts-Austen, William 132
robots 73, 241
Rockwell hardness tester 101
Rodin, Auguste 70
rolling 80
Rolls-Royce 110, 157, 273
Romans 2, 183
Royal Mint 93, 95
runners (in casting) 66
Russia *see* USSR, former
ruthenium 217
Rutherford, Ernest 251

sacrificial protection 230
safety 290
samarium 216
sand casting 66
sand for moulds 68
scandium 218
scanning Auger microscope 55
scanning electron microscopes 54
scrap metal 116, 122, 183, 274
sea
   as source of energy 280
   as source of metals 197, 267
seam welding 238
'season cracking' 231
Sefström, Nils 224
selenium 218
self-lubricating bearings 248
semiconductors 5, 205
semimetals *see* metalloids
semi-solid alloys 76
Sendzimir mill 83
sequence casting 125
shape-memory alloys 271

sheet metal   89
shell moulding   68
sherardizing   181
ships   114, 161, 162, 235, 240
SI units   xvi, 98
Siemens, Charles and Frederick   116
silica (SiO₂)   17, 21
silicon   218
  in aluminium alloys   155
  amorphous   279
  in cast iron   147
  chips   48, 219
  in steel   218
silicon nitride   286
silver   94, 220
silver solder   236
single crystals   48, 219
sink-and-float process   275
sintered metal powders   247
sintered ores   22
slag   21
slip, in crystals   62
smelting   8, 18
Soddy, Frederick   250
sodium   155, 220, 253, 263
sodium–potassium alloy   198, 215
solar energy   279
solders   38, 177, 185, 235
solid solutions   49
solution heat treatment   159
solvent extraction   16, 188, 207
Sorby, Henry   4, 132
South Africa   13, 217
space frames   37, 165
space vehicles   206, 207, 208
spark plugs   37, 208
specifications   155
spent pot linings (SPL)   35
spheroidal graphite iron   148
spot welding   238
squeeze casting   76
stained glass windows   185
stainless steel   94, 141, 189, 228, 257,
     264, 293
Statue of Liberty   115, 235
steels   97, 113
  alloy   140
  composition   113

continuous cast   123
  heat treatment   133
  metallography   128, 130
  production   115
  products   129
  rolling   80
  tubes   84
  see also die steels, high speed steels,
     HSLA steels, manganese steel,
     maraging steels, nickel steels,
     stainless steel, superalloy steels
Stellite alloys   205
sterling silver   220
stoves   24
stranger atom   59
Strassmann, Fritz   253
street lighting   203, 212, 221
stress–corrosion   230
strontium   221
sulphur, in iron and steel   113, 122, 209
superalloys   48, 192
superalloy steels   145
superconductivity   212, 293
Superfund Sites   35
superglues   244
superplastic alloys   42, 199
surface treatments   136
surgical uses of metals   136, 199, 221
Sweden   279
Switzerland   31
swords   237
Système Internationale d'Unités   xvi, 98

Tacoma Narrows Bridge   292
tantalum   221
Tay Bridge disaster   116
Taylor, Fred   144
technetium   222
tectonic plates   267
telescopes   154
tellurium   222
temper brittleness   140
tempering   134
tensile tests   98
terbium   216
Thames Barrier   229
thermocouples   217
thermonuclear fusion   209

thermoplastics 282
thermosetting plastics 283
thorium 222
thulium 216
timber 281
tin 175
  alloys 177
  ores 14
  properties 175
tinplate 175
tipped tools 200, 223
titanium 13, 145, 198
tool steels 200
Tour de France 196, 287
Toyota deposition process 139
transistors 205
Troy weight 206
tube-making 84
tungsten 222, 268
tungsten carbide 144, 222
tungsten inert gas welding 239
Turner, Thomas 147
Tutankhamun 206
tuyères 23, 25
type metal 183

ultrasonic testing 103
ultrasonic welding 238
undersea welding 242
university training 289
uranium 223, 253, 264, 267
uranium enrichment 254
uranium hexafluoride 254
USA 33, 35, 83, 126, 165, 183, 196,
  211, 224, 255, 280, 291, 294
USSR (former) 19, 169, 217

vacancies 62
vacuum degassing 122
vacuum diecasting 76
vacuum heat treatment 136
vacuum melting 145
vanadium 145, 224

verdigris 8, 169
vertical retort process 179
vertical serpentine process 176
Vickers hardness tester 101
Victoria Cross 173
Volkswagen 195
Voyager spacecraft 208

Washington, H. S. 8
Wedepohl, Karl 10
welding 237
Wheatstone, Charles 170
White, Maunsel 144
Wilm, Alfred 157
wind as source of energy 280
wire drawing 88
wire feed injection 123
Wöhler, August 104
Wöhler, Friedrich 30
Wollaston, William 217
Worshipful Company of
  Goldsmiths 207
wrought iron 17, 44, 114

Xerox copiers 218
X-rays 56, 102, 185

Young, Thomas 300
ytterbium 216
yttrium 216, 225

Zaire 268
Zamak 182
Zambia 241, 268
zinc 178
  alloys 74, 181
  coatings 180
  ores 178
  production 178
  uses 180
zirconium 195, 225, 260, 265
zone refining 219, 300

Discover more about our forthcoming books through Penguin's FREE newspaper...

It's packed with:

- exciting features
- author interviews
- previews & reviews
- books from your favourite films & TV series
- exclusive competitions & much, much more...

Write off for your free copy today to:
Dept JC
Penguin Books Ltd
FREEPOST
West Drayton
Middlesex
UB7 0BR
NO STAMP REQUIRED

READ MORE IN PENGUIN

In every corner of the world, on every subject under the sun, Penguin represents quality and variety – the very best in publishing today.

For complete information about books available from Penguin – including Puffins, Penguin Classics and Arkana – and how to order them, write to us at the appropriate address below. Please note that for copyright reasons the selection of books varies from country to country.

In the United Kingdom: Please write to *Dept. JC, Penguin Books Ltd, FREEPOST, West Drayton, Middlesex UB7 0BR*

If you have any difficulty in obtaining a title, please send your order with the correct money, plus ten per cent for postage and packaging, to *PO Box No. 11, West Drayton, Middlesex UB7 0BR*

In the United States: Please write to *Penguin USA Inc., 375 Hudson Street, New York, NY 10014*

In Canada: Please write to *Penguin Books Canada Ltd, 10 Alcorn Avenue, Suite 300, Toronto, Ontario M4V 3B2*

In Australia: Please write to *Penguin Books Australia Ltd, 487 Maroondah Highway, Ringwood, Victoria 3134*

In New Zealand: Please write to *Penguin Books (NZ) Ltd,182–190 Wairau Road, Private Bag, Takapuna, Auckland 9*

In India: Please write to *Penguin Books India Pvt Ltd, 706 Eros Apartments, 56 Nehru Place, New Delhi 110 019*

In the Netherlands: Please write to *Penguin Books Netherlands B.V., Keizersgracht 231 NL–1016 DV Amsterdam*

In Germany: Please write to *Penguin Books Deutschländ GmbH, Friedrichstrasse 10–12, W–6000 Frankfurt/Main 1*

In Spain: Please write to *Penguin Books S. A., C. San Bernardo 117–6° E–28015 Madrid*

In Italy: Please write to *Penguin Italia s.r.l., Via Felice Casati 20, I–20124 Milano*

In France: Please write to *Penguin France S. A., 17 rue Lejeune, F–31000 Toulouse*

In Japan: Please write to *Penguin Books Japan, Ishikiribashi Building, 2–5–4, Suido, Tokyo 112*

In Greece: Please write to *Penguin Hellas Ltd, Dimocritou 3, GR–106 71 Athens*

In South Africa: Please write to *Longman Penguin Southern Africa (Pty) Ltd, Private Bag X08, Bertsham 2013*

READ MORE IN PENGUIN

SCIENCE AND MATHEMATICS

QED   Richard Feynman
The Strange Theory of Light and Matter

'Physics Nobelist Feynman simply cannot help being original. In this quirky, fascinating book, he explains to laymen the quantum theory of light – a theory to which he made decisive contributions' – *New Yorker*

Does God Play Dice?   Ian Stewart
The New Mathematics of Chaos

To cope with the truth of a chaotic world, pioneering mathematicians have developed chaos theory. *Does God Play Dice?* makes accessible the basic principles and many practical applications of one of the most extraordinary – and mind-bending – breakthroughs in recent years.

Bully for Brontosaurus   Stephen Jay Gould

'He fossicks through history, here and there picking up a bone, an imprint, a fossil dropping and, from these, tries to reconstruct the past afresh in all its messy ambiguity. It's the droppings that provide the freshness: he's as likely to quote from Mark Twain or Joe DiMaggio as from Lamarck or Lavoisier' – *Guardian*

The Blind Watchmaker   Richard Dawkins

'An enchantingly witty and persuasive neo-Darwinist attack on the anti-evolutionists, pleasurably intelligible to the scientifically illiterate' – Hermione Lee in the *Observer* Books of the Year

The Making of the Atomic Bomb   Richard Rhodes

'Rhodes handles his rich trove of material with the skill of a master novelist ... his portraits of the leading figures are three-dimensional and penetrating ... the sheer momentum of the narrative is breathtaking ... a book to read and to read again' – Walter C. Patterson in the *Guardian*

Asimov's New Guide to Science   Isaac Asimov

A classic work brought up to date – far and away the best one-volume survey of all the physical and biological sciences.

READ MORE IN PENGUIN

SCIENCE AND MATHEMATICS

The Panda's Thumb   Stephen Jay Gould

More reflections on natural history from the author of *Ever Since Darwin*. 'A quirky and provocative exploration of the nature of evolution ... wonderfully entertaining' – *Sunday Telegraph*

Einstein's Universe   Nigel Calder

'A valuable contribution to the demystification of relativity' – *Nature*. 'A must' – *Irish Times*. 'Consistently illuminating' – *Evening Standard*

Gödel, Escher, Bach: An Eternal Golden Braid
Douglas F. Hofstadter

'Every few decades an unknown author brings out a book of such depth, clarity, range, wit, beauty and originality that it is recognized at once as a major literary event' – Martin Gardner. 'Leaves you feeling you have had a first-class workout in the best mental gymnasium in town' – *New Statesman*

The Double Helix   James D. Watson

Watson's vivid and outspoken account of how he and Crick discovered the structure of DNA (and won themselves a Nobel Prize) – one of the greatest scientific achievements of the century.

The Quantum World   J. C. Polkinghorne

Quantum mechanics has revolutionized our views about the structure of the physical world – yet after more than fifty years it remains controversial. This 'delightful book' (*The Times Educational Supplement*) succeeds superbly in rendering an important and complex debate both clear and fascinating.

Mathematical Circus   Martin Gardner

A mind-bending collection of puzzles and paradoxes, games and diversions from the undisputed master of recreational mathematics.

READ MORE IN PENGUIN

SCIENCE AND MATHEMATICS

The Dying Universe Paul Davies

In this enthralling book the author of *God and the New Physics* tells how, from the instant of its fiery origin in a big bang, the universe has been running down. With clarity and panache Paul Davies introduces the reader to a mind-boggling array of cosmic exotica to help chart the cosmic apocalypse.

The Newtonian Casino Thomas A. Bass

'The story's appeal lies in its romantic obsessions ... Post-hippie computer freaks develop a system to beat the System, and take on Las Vegas to heroic and thrilling effect' – *The Times*

Wonderful Life Stephen Jay Gould

'He weaves together three extraordinary themes – one palaeontological, one human, one theoretical and historical – as he discusses the discovery of the Burgess Shale, with its amazing, wonderfully preserved fossils – a time-capsule of the early Cambrian seas' – *Mail on Sunday*

The New Scientist Guide to Chaos edited by Nina Hall

In this collection of incisive reports, acknowledged experts such as Ian Stewart, Robert May and Benoit Mandelbrot draw on the latest research to explain the roots of chaos in modern mathematics and physics.

Innumeracy John Allen Paulos

'An engaging compilation of anecdotes and observations about those circumstances in which a very simple piece of mathematical insight can save an awful lot of futility' – Ian Stewart in *The Times Educational Supplement*

Fractals Hans Lauwerier

The extraordinary visual beauty of fractal images and their applications in chaos theory have made these endlessly repeating geometric figures widely familiar. This invaluable new book makes clear the basic mathematics of fractals; it will also teach people with computers how to make fractals themselves.